For a free complementary copy of the book,
Omnipresent who is IT,
And its Translation, and more, write to
omnipresentrr@gmail.com

Preface

Throughout this book you will find the frequent use of the word ***IT***, which will be in bold, capitalized, italic letters, as a neutral pronoun. The meaning of ***IT*** unfolds with the progression of each section, yet until the end, ***IT*** may not be altogether clear, so until then, consider ***IT*** as a new way to say God, Creator, or even "pure energy." ***IT*** signifies that which put and keeps this Universe together.

Additionally, this book contains information intended to challenge your current understanding of the Universe and its workings along with our existence within ***IT***. The topics covered may be familiar, and they are backed by science or personal experience, yet it is unlikely that you have viewed our existence from the perspective of Omnipresent. To fully understand the atom, for example, or to appreciate the production of garbage or the reshaping of ***IT,*** you would benefit from preparing for this book as you would a meditation - by unlocking the mind.

To assist you in unlocking your mind I have included the photon that follows. Look at it, and think about what you see.

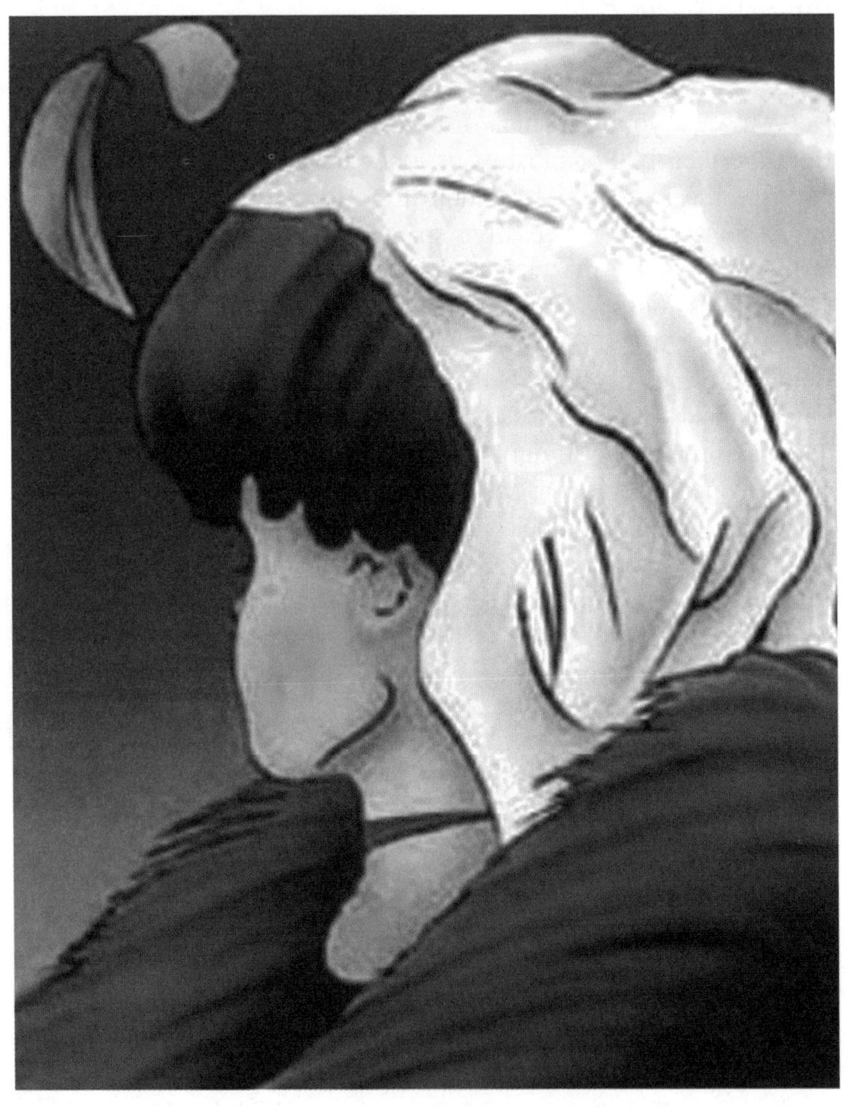

[1]This artwork is in the public domain. It was first publish as an anonymous German postcard ca. 1888. It was redone for an advertisement for the Anchor Buggy Company from 1890. Thereafter, British cartoonist W. E. Hill published it in 1915 in *Puck* humor magazine, an American magazine inspired by the British magazine *Punch*.

British magazine *Punch*.

Do you see a young woman or an old woman? Can you see both? Both are there! For most of us, once our mind locks in one view, usually the first, it will seldom look for alternatives. Our first impressions come so easily, so quickly, yet considering how influenced we are by what we perceive, we should be aware that first! Impressions are limited, incomplete, and very often incorrect.

Being informed that more than one figure in the picture exists likely makes it easier to see the image differently. However, the question remains: Without knowing an extreme alternative existed, would you have tried to see it? By first accepting that alternatives exist, you begin to unlock your mind, clearing it of all the first impressions enabling you to see the other image that is also there. Subsequently when we take the opportunity to view even old ideas from new angles, we will find those quick and easy ways of seeing are not the right or best methods to use when engaging our minds in a new thought.

Part #1 of Omnipresent sets forth what I call trivia: a collection of thoughts meant to provide alternative ways of seeing and being. It's a primer for sections two and three, which uncover questions and truths about time, matter, and the mass of the universe, about God, and ***IT*** as an ever changing place of omnipresence. I hope you will enjoy the book. How I came upon the material within is a story in itself, but in summary I can say it all stems from gratitude. I am grateful that I exist and that I accept and question what I see. In doing so, I have seen as most of us see. And because of my gratitude I have learned to see alternatives, which has helped me in a personally profound way to know who I really am.

Introduction

In this section we will be talking about matter and speed. More than that, we will be talking about all the emptiness that exists as the Universe and inside every atom. This section will help you to accept that everything that exists as pure energy, as God, and as ***IT,*** is, and is omnipresence.

For some this will be easy reading, and for others it won't be. But this is an all-in-one book put together for those who do not understand certain subjects. I hope you re-read the parts that make no sense to you. I hope all readers read this slowly and more than once so that what was already clear may act as a light to see the more shadowy parts. You will likely want to rethink some things that are ordinarily thought of as normal. Try to forget the existing labels that we have placed on this Universe and the way we believe that it should be.

Please take this information as something that will help you understand ***IT*** - pure energy- the creator-God-that which makes everything possible. Do not even look at this text as mere information but rather as an example of ***IT***-pure energy-, reshaped as matter and as weight as ***IT*** plays with speed controlled by electrons. For in this section, we will see, using simple mathematics, that when we take away ***ITS*** weight, this pure energy, as a duality, will add up to 99.99% nothingness as ***ITSELF***, as pure energy.

We can now, with advanced information and technology, understand this nothingness, as a part of that which is called Dark Matter, even though it is not really dark; it is clear, by which I mean transparent. You will read how this emptiness exists in everything, including our bodies. You will also read why light is not an entity itself but a by-product, and here too is why time does not stop at the speed of light, but rather that light can move, for example, from the Sun to other parts of ***ITSELF*** behaving as a wave and a particle simultaneously.

First however, we shall look at the Big Bang as pure energy: ***ITS***

weight and density, and ***ITS*** spin and speed of 186,000 miles per second (186k mps). We will look at water and our Universe and the three scenarios of ***ITS*** existence including ***ITS*** constant presence, ***ITS*** omnipresence, as nothingness. This is not a Science 101 class, but rather information on ***IT***, which we have been able to gather with the aid of modern science and technology.

Let me also mention to you, the reader, that, in general, we see things as having mass, and this has to be so, because, naturally, in order to see anything we do need to see it as the mass that ***IT*** exist as.

Now let me explain something that will help you understand some of the subjects that you will read about in this section, and that you will better understand what I am writing about, if you remember that for me to understand ***IT*** better, I have had to see ***IT*** as ***IT*** is now, and how ***IT*** existed at the moment of the Big Bang, and even before. But to do this I have had to reduce my perception of the many ways that ***IT*** exists now, that would still be the same, now, and the same at the moment of the Big Bang, and how ***IT*** could have existed before the Big Bang. So please remember that it is only a way of seeing and understanding the way ***IT*** exists that I am presenting to you in what you will be reading, as I see ***IT,*** as ***ITS*** duality: first as ***IT***S nothingness, and second as ***ITS*** weight. Furthermore, ***IT*** is from ***ITS*** weight that anything that can exist, exists, and **anything that exists has to exist within *ITSELF.*** I would also like my readers to be aware that I am not a chemist, or a physicist, and as to what my education is composed of, I will elaborate more later on in this book.

But returning to my main point, I want you, the reader, to bear in mind that everything that I have written in this section and what follows is based on a question I asked of ***IT.*** I simply asked ***IT*** if ***IT*** would allow me to understand ***IT*** better.

Now this does not mean that everything that I have stated herein is an absolute truth, for I too have my limitations. For this reason I welcome all points of view that relate to ***IT***, for I too know that

there is a lot more to learn about ***IT***, that other readers have, in terms of information, and to me, ***IT*** is not important how ***IT*** exists outside of me, as much as how ***IT*** exists within me. I know that ***IT*** is huge, and that we are just beginning to understand ***IT*** better, and to me, no matter how ***IT*** exists outside of me, in ***ITS*** peaceful ways, and as ***ITS*** violent ways simultaneously, I will always have to say to ***IT***, thank you for allowing me to be here in observing the way ***IT*** exists. One thing I have learned is that ***IT*** did have weight at the moment of the Big Bang, and this weight still exists as the weight of all the matter that now exists within this Universe. This very same weight has always existed inside of ***ITSELF***, which is made of a cold, clear, invisible nothingness, as ***ITS*** body. For this reason, when I say that light must have an attached weight it is because this light is something. Everything that is something must have weight; ***ITS*** weight.

Another thing for you, the reader, to consider is that there is a law in physics that states that if something cannot be disproved, then it is possible. But I think that what is more important is that what I have offered you in this section will help you understand ***IT*** better, not in terms of physics or chemistry, but in terms of how ***IT*** exists as omnipresent.

Clearness as primary

I believe clearness or transparency may be primary to ***ITS*** existence. We know that colors exist in their various ways, first, because seeing is believing, and we can visually see all the colors that exist in our environment from the colors in minerals to the colors we apply as paint on objects. Second, we know scientifically that color exists in light as all the colors of the spectrum. But to have color from minerals, we return again to atoms. Without atoms, we do not have minerals or the heated dense matter that ***IT*** exists as. And in order to have colors, the heat of atoms that produces minerals has to exist. Hence, without the part of ***IT*** that exists as very heated dense matter, we cannot have color. And no, I have not forgotten about the colors that exist in the light spectrum. Here too, the colors that exist are only possible because of light, which on our Earth is a by-product of the Sun, which is composed of atoms of hydrogen and helium. So here again, colors come from that part of ***ITSELF*** that exists due to ***ITS*** heated weight, the same heated weight that forms matter.

But when ***IT*** existed prior to the Big Bang, ***IT*** had not yet reshaped into atoms to produce the minerals that would have color as an after effect. Before the Big Bang there were no stars to produce light or heat or subsequent matter, which is what now gives us the colors from yellow to red to blue. So this clearness does not come from light. In the pre-Big Bang stage of ***ITS*** existence, there were no colors as we now know them. This is significant, because what existed then still exists now. Clearness, or transparency, is the primary form of ***IT*** that exists in ***ITS*** omnipresence. Remember, as ***IT*** existed before the Big Bang, ***IT*** had all its weight as a single unit and the same nothingness that now exists is also the same nothingness that existed then. ***IT*** is all that ***IT*** was and has ever been as a clear, transparent way of existing, just a clear something; not a color or a light, for both of these came after the Big Bang. ***ITS*** colors are a by-product of its weight from atoms, not to exclude light, which is also a by-product, so ***IT*** has to exist primarily as a clear, transparent nothingness.

Furthermore, ***ITS*** clearness or transparency allows us to see through ***ITS*** nothingness. If it were not for this clearness we would not be able to see the colors that now exist due to ***ITS*** heat.

Clearness is a natural quality. When you shine light on it, it remains transparent, and when you remove light from it, it becomes dim, appearing darker. **Darkness and light are not primary characteristics, because darkness is the absence of light, and light is a by-product that exists as a result of *ITS* reshaping *ITS* weight into stars.**

Dark Matter = clear nothingness
Dark matter is considered to be that area of the Universe that seems to be empty and where no light from any source penetrates. It is that 95% empty space of the Universe which I later refer to as being 99.99% nothingness. As far as I am concerned, we call it Dark Matter because it is there, and we figure that since it exists, it needs to exist as something. But why can't it exist as nothing?

We need to remember that Dark Matter is not like the matter we know, but it is a freezing cold substance. It has substance in the actual space it occupies. The fact that it is different from the matter we know and can see does not mean that it does not exist.

It appears dark to us and to the astronauts when they are out there; we do not have light to penetrate it. In our case, light is only a by-product of the Sun; but light comes from many sources other than our Sun. Still, we can penetrate the absolute darkness of the majority of space.

Without sunlight we die. When the Sun dies, this area will also become part of the vast darkness. Let us not dwell on this; there is more to it. When you see anything at any distance, it is not because there is nothing between you and that object. There are billions of atoms that could obstruct your vision. But most atoms are clear in their composition, so you can see through them.

The first layer of any atom contains the electrons, through which we can see. And each electron has 95% empty space, through which we can also see. We also see through the center of the atoms, where the protons and the neutrons are; they are also clear

or transparent because light passes through them without obstruction. On the other hand, opaque substances do not allow light to pass through them freely, but instead absorb and reflect specific wavelengths of the light spectrum. We perceive things as being of a certain color because our eyes detect the wavelength of light that the object reflects. In a sense we could say that colors obstruct our vision.

Dark Matter II

Imagine that you are in a dark closet. It's pitch black. You see absolutely nothing. You do not know where anything is. Let us say that you have the light dimmer in your hand and you gradually turn it up. You start seeing things again. Well, imagine if you could bring the light up throughout the Universe. What was impenetrable darkness would be shown to be completely clear: a lot of empty, freezing cold space through which we can see. Dark matter is the clear, empty nothingness that ***IT*** exists as.

My thought on why there is so much darkness is that if there were much more sunlight, it would be too hot to allow life to exist.

Experiment for Astronauts: S.O.S. Dark Matter

To confirm the truth about Dark Matter, astronauts might take a trip to outer space, going as far from the Sun as they can, and fill a clear, hermetically sealed container with Dark Matter and bring it back to Earth. Then we could study it. As I previously stated, I think we will find that it is not dark but clear. However, it would be difficult to determine this because any kind of material environment, including a lit one, could change the nature of what was captured within the sealed container.

How do you capture Dark Matter, which is quite possibly absolute nothingness and study it? How do we write and talk about, not to mention study and measure, absolute nothingness? Is it the opposite of what we know as matter? Will we be able to use Dark Matter as fuel? Is this beyond the atom? Could we use the dynamics of absolute nothingness to fuel our future travels into space?

When humanity was primitive, it had no idea of how to use any of the invisible things that we now use for energy; like gases. We

have been using oxygen to breathe all along, but now it fuels many other things besides our bodies. Isn't it possible we could do the same with Dark Matter?

It might take thousands of Earth's rotations before we find the way to use Dark Matter. But there must be a way to channel all that speed of absolute, cold emptiness. There is so much of it!

We should at least be able to find our duality. Out of this fixed planet, we should step into the huge nothingness and not forget that we once thought our puny world was just as huge.

One thing is for sure: Space travelers will not have to worry about refrigeration; it is cold out there. Their problem will not be eating cold ice cream but how to have a warm meal. Of course, in the future we probably won't even need warm meals.

The cold of deep space will be useful for all kinds of cryogenic issues: suspended animation for long term travelers, food storage, storage of dead bodies, etc. There will be a lot of business going on in outer space; the cold out there needs to be harnessed to our advantage. As we venture out, ***IT*** will be searching for all possibilities. For now, all we can do is to continue to learn and understand the nature of the Universe as omnipresent. And we should be grateful for existing in this omnipresence.

The glass ball analogy

In trying to find ways to describe ***IT*** better, I find this might come close: Take one of those round picture balls or snow globes with scenery inside a clear liquid. If you shake it, the scenery moves, but it stays contained within ***ITSELF***. We can see this as the concept of the conservation of pure energy; everything contained within--nothing created or destroyed. We can thus also see this as a model of ***ITSELF*** contained throughout the Universe but with the difference that ***IT*** is always changing, as ***IT*** reshapes ***ITS*** weight.

Imagine that this clear ball is the Universe and because of the light, all is clear; we can see everything. The clear part is the cold emptiness that ***IT*** exists as. Of course, in the picture ball all the particles remain static while the real Universe is constantly changing from the subatomic level to all the matter that exists in

the whole Universe. This gives us transmutation for the reshaping of ***IT***.

Because of the constant reshaping, we have been fooled into thinking there is a past, present, and future. These things cannot be unless we use the concept of time, which we know is not universally applicable. The movement of our planet also fools us. We see the rotation of the planets and movement in general as intricately related to time and distance.

To understand motion, all we have to do is go back to the picture/scenery ball and shake it. In shaking the ball, you are applying energy to produce motion. In the case of the Universe, the motion was provided by the Big Bang. Since then, ***IT*** has been moving in an outward motion.

Imagine the microscope you need to see every particle inside the picture ball. Now imagine the opposite of that, the telescope, that you need to see the real Universe. Without the telescope, I can barely see the Moon. As we look out we are not even sure where we are in the totality of ***IT***. We are however learning more with the Mars rovers and the Cassini probe. Wonderful technology! Think of the immensity!

Let me add another way of viewing this crystal ball. Imagine that the crystal ball you are viewing has no outside shell like a crystal ball has, for we should remember that the Universe, as ***ITS*** nothingness, does not have an external shell. As you view this crystal ball without the shell what will happen is that you will see ***ITS*** weight as celestial bodies moving within. But this crystal ball does not have a beginning or end, and as you look you will see darkness because there is no light inside it. Yes, you can see inside because there are stars that are lighting up the surrounding areas in this stage of ***ITS*** existence.

If we could view ***IT*** in stage 1, pre-Big Bang, what you would see is this possibility of what ***IT*** might look like. We might find it difficult to visualize. Because we cannot exist in that stage of ***ITS*** existence, with all of ***ITS*** weight concentrated in one point, we

have to use our imagination to see these various possibilities, as if we were looking at ***IT*** as a crystal ball that does not have a shell.

The first problem will be that since ***IT*** exists as its own nothingness, as a cold, clear nothingness, I will not be able to see ***IT*** internally or externally, but I will be able to detect its coldness. Somewhere inside this nothingness there has to be ***ITS*** very heated (trillions of degrees hot), dense weight, for this heated weight, as pure energy, has to continue existing somewhere inside of ***ITSELF***.

Let me add that as we imagine this situation, the existing Universe, as the coldness and as the heat that now exists as pure energy, would also have had to exist before the Big Bang.

So, getting back to one possibility of what ***IT*** might exist as, it would be a very strange way of existing, and here is why: The coldness that now exists as pure energy would also have had to exist before the Big Bang. This is strange because ***ITS*** heated weight must exist within its coldness simultaneously; both existing in harmony.

I would like to add one more possibility to the above scenario. Since ***ITS*** outside is cold and clear, what is related to ***ITS*** heated weight might be clear also. I say "might" because we, as humans, will never be able to be present in the pre-Big Bang stage of ***ITS*** existence.

Another reason we might not be able to see ***ITS*** weight is that a lot of it is clear (that is, transparent). As I look at things outside of me, I see them because I can see through ***ITS*** weight. The best example is seeing through the air that surrounds us, this is the weight that exists in the protons and the neutrons of the oxygen atoms, and in the atoms of the other gases that make up our atmosphere. So, it might be that we will never be able to see ***ITS*** weight in pre-Big Bang form, especially if so much of it is similar to air, and other gases.

But again, what we are doing is exercising our minds, imagining how ***IT*** could exist as different possibilities. What is important is

to always return to the way ***IT*** exists as you and I and everything else within this omnipresence.

ITS being clear
Here is one more way to see ***IT*** existing as a clear or transparent form of being:

Since the Universe is composed of 90% hydrogen atoms, this means that as we view the Universe from where we are, looking into ***IT*** as a whole, that is, when we see all the celestial bodies that exist out there, we can see straight through ***IT*** without obstruction. Hydrogen is transparent, and when we see through these hydrogen atoms, we are also seeing through the hydrogen atom electrons and the empty space that exists inside the atom and straight through the one proton that makes up the nucleus of each hydrogen atom. So you see, if ***IT*** were not a clear form of being, we would not be able to see straight through what ***IT*** now exists as: ***ITS*** huge invisible housing or body, where ***IT*** exists as being 90% hydrogen atoms.

Maybe the above situation will be clearer to you if you remember that everything that exists is ***IT***, as pure energy, or as God. Now this 90% that is made up of hydrogen atoms has to exist somewhere, and that somewhere is inside of ***ITSELF***, as what we see and understand as this Universe.

It is in ***ITS*** bringing together ***ITS*** duality that we get light, and ***IT*** is in light that we have all the colors that can exist, but not independently as in ***ITS*** nothingness or ***ITS*** heated weight, because when we see through ***ITS*** nothingness, (that is, what we perceive as the Universe) there are no colors, and we can see straight through ***ITS*** heated weight, as when we see through the protons that exist in hydrogen atoms and oxygen atoms.

If you find this difficult to understand, try to see ***IT*** this way: In order to see colors you do need the presence of light, and light is a positive and negative coming together as ***ITSELF,*** as a duality.

⌘~~~~~~~~~~~~~~~~⌘⌘~~~~~~~~~~~~~~~~⌘

******IT is ITS nothingness that is always alive and awake, as IT watches what IT does with ITS weight *******

IT as all-knowing

When I look up at the sky and into ***IT*** as the Universe at night, I am aware that everything out there, and everything inside of me, exists as ***ITS*** oneness. If I ask myself how every atom in my body knows the changes it has to make, or how a planet or star that may find itself at the other end of this Universe knows how it is supposed to make changes, or how everything that exists knows what it, as weight, is supposed to change into, it helps to remember that ***IT*** is fragmented into different portions of ***ITSELF*** as weight because the weight I exist as is not the same weight that you exist as; everything exists as different fragments of ***ITS*** weight so that my weight cannot, to use a word, "communicate" with your weight, and even less, planets at opposite sides of ***ITSELF***. Yet ***IT*** is all-knowing, and is in all places at the same moment, known as omnipresent. While ***ITS*** weight does not exist as a oneness in this stage of ***ITS*** existence, ***ITS*** weight exists within ***ITSELF*** as the oneness that ***IT*** exists. Therefore, what we see as matter is just ***ITS*** fragmented weight that exists within ***ITSELF***.

The only thing that exists as one and is everywhere as the same moment, that is not fragmented and is constant, is this divine consciousness that is the bulk nothingness that IT exists as.

Humans exist as fragments of ***ITS*** weight. This fragmented human body that we exist as, as weight or as an object, is where ***IT*** exists as the operator, for I can assure you that I do not run or operate this human body that I exist as.

So now see if you can think about this in ***ITS*** reversed form, as the weightless Universe where ***IT*** exists as a body and where ***IT*** is operating as the mover and shaker of ***ITS*** interior weight within ***ITSELF***.

When I think of what ***IT*** is that can communicate with every atom in my body, in terms of what this atom (weight) is supposed to do, change into, or evolve as, and for how long, and for how many repetitions, the only thing I can find that could do this is the same nothingness that exists in between an atom's electrons (speed) and protons and neutrons (weight), for this nothingness that exists in all

atoms is the same nothingness that exists as this divine conscious nothingness that ***IT*** exists as; that I call it ***ITS*** shell or housing.

When we look at ***ITS*** weight, it will always exist within ***ITSELF*** as a shell of Dark Matter, the nothingness of the Universe. As this nothingness is all-knowing, every atom gets its instructions from ***ITSELF*** as ***ITSELF***, and within ***ITSELF*** as the only operator, for there are no other gods in this Universe that exist as this Universe.

So it is easy to understand that ***IT*** is all-knowing if you remember that ***IT*** is, always was, and will be, ***ITSELF***--not as a male or female, but rather as something that is a divine form of conscious nothingness that has been reshaping ***ITS*** weight within ***ITSELF*** so as to allow us to come into contact with ***ITSELF*** as human minds which exist as ***ITS*** own weight, in the form of our human bodies, so that we can see the many changes that ***IT*** has gone through as ***ITS*** weight, which is what we have come to understand as the history of ***ITS*** evolution, and that now, we will be seeing ***IT*** as ***IT*** reshapes into what we are calling high-tech technology, for we are now going to understand more about who ***IT*** is in ways that we could not have understood when we first arrived and existed as primitives on this part of ***ITS*** weight called Planet Earth.

IT as a spirited nothingness

Now, I have been using words to describe ***IT***, as to how it basically exists, like ***IT*** being a shell, housing, or body. But this is only so you can understand that what I am trying to say related to ***ITS*** way of existing. These words all have one thing in common: They describe something physical. In reality, however, to use a word to describe ***IT*** as a mass or ***IT*** as a body is misleading, for ***IT*** does not exist as something that is made of matter. ***IT*** basically exists to us as a nothingness, and the only thing that I can think of as a word in current usage to describe ***ITS*** existence is the word spirit, for the word spirit is supposed to denote something that exists, but is not made of physical matter.

One thing we can use as something that does exist is ***ITS*** size. For us to understand how big ***IT*** is, we can begin to use what we know as distance as a marker, that ***IT*** exists as, and as the matter that

now exists in what we see as the distance from one galaxy to another in this Universe.

I am using this example because now we know that all ***ITS*** fragmented weight exists within ***ITS*** huge nothingness as size, in terms of the way we have focused on ***IT***, as the matter that does exist in a place called the Universe, but not as ***God***.

We know that a Universe does exist, and that within this Universe there is matter, as planets, stars, and galaxies, and that there is distance between one galaxy and another, and that these galaxies exist in a place. Here we should take a moment to remember that we have been conditioned to see things that exist as a way to understand our reality, when in fact, what we see is an illusion. For example, we see days as being new only because of the rotation of our planet, and we see things as physically being there, as solids, when in reality we are seeing something that is at least 95% nothingness.

As we move from planet to planet, we are traveling distances within ***ITS*** nothingness. So when we look at the Universe, we are primarily focusing on things because we can see them. But we should remember that what we are seeing as ***ITS*** fragmented weight is also an illusion, because what we are seeing as ***ITS*** weight (matter = stars and planets that are made of atoms, which are also 95% nothingness) has to exist within the rest of ***ITSELF***, as the hugeness that this Universe exists as. This is the best illusion that the human mind can toy with because if this illusion did not exist, neither would we as the greatest illusion: the human body and the human mind, which are made of this nothingness, as the mass (weight) that we exist as in a human body, so that as ***ITS*** weight we can, as human minds, feel that we exist independent of ***ITSELF***.

It has helped me to understand ***IT*** better if I remember that I as RIC may feel as though I am one entity, but my being one is only so because I exist as the trillions or more fragments of ***ITS*** weight that exist within ***ITS*** conscious nothingness. I say this, because ***IT*** also exists as one, and as one, ***IT*** is just moving ***ITS*** weight around within ***ITSELF***, inside what we see as a Universe, the same way

that we will be seeing the things that will happen in nanotechnology, as being a world in itself. But to me, I know that even this nanoworld could not exist if ***IT*** did not exist.

And with respect to ***ITS*** hugeness, it is relative to ***IT***, for ***IT*** does not have anything to compare ***ITSELF*** to as far as ***ITS*** size goes. To us ***IT*** is huge because of the distance that exists from one galaxy to another in terms of matter, which to ***IT*** is just ***ITS*** weight within ***ITS*** nothingness.

And as a cold, conscious nothingness, ***IT*** can think, if one can use the word think, for how else can something be so perfect in terms of the things that ***IT*** can reshape into by just using a billionth to a trillionth of ***ITSELF*** as weight within ***ITSELF***, so that stars and planets and humans can exist? If you think that it is strange, or weird, as to the way ***IT*** has used ***ITS*** weight to reshape into stars and planets, keep an eye on how ***IT*** is going to use an even lesser fraction of ***ITS*** weight as ***IT*** continues reshaping in this new area known as nanotechnology. However, one thing that we should remember as we watch this nanotechnology area, is that ***IT*** will continue to exist in this very tiny area, where we will see motors, switches, and what not, which would not be able to exist if it were not for the nothingness that also has to exist in this miniature world that ***IT*** will construct, which will also include ***ITS*** qualities of speed and spin.

And as for me, I have to be thankful for the effect that ***IT*** produces as an illusion because now I see a person in front of me as the effect of ***ITS*** reshaping into that human that is made up of atoms that are 95% nothingness. And as they talk I understand that they can think and talk only because of the emptiness that exists as the atoms which compose their brains. None of these empty atoms could give the person in front of me a chance to be conscious if it were not for the consciousness that ***IT*** exists as in the form of a divine life as a conscious nothingness.

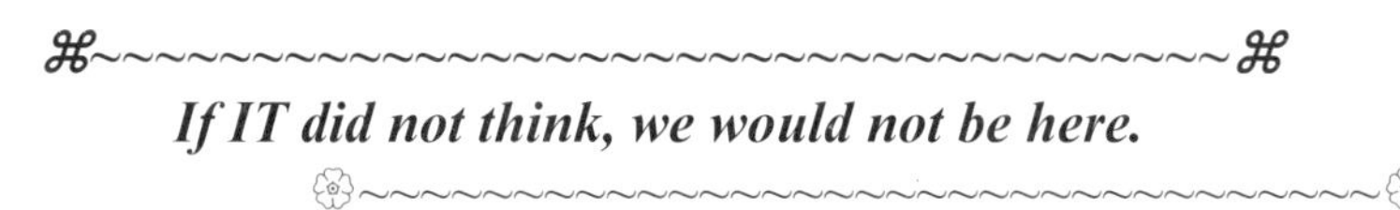

If IT did not think, we would not be here.

Omnipresent and God as nothingness

To understand ***IT*** better, I have added two words: God and omnipresent. First, the word Omnipresent refers to that which is in all places at the same moment. As applied to God, omnipresent means God is in everything that exists as matter, including the nothingness that exists inside of matter. This would also include planets, stars, galaxies, and everything else that may exist including Dark Matter itself, and the nothingness that exists in every atom. How else could God be everywhere at the same moment if ***IT*** were not for everything that exists in this 99.99% nothingness?

And the second word is the word God, which many use in reference to the one who created every thing that now exists. But God didn't create anything that didn't already exist as God. God is everything that exists as matter: planets, stars, and what the mind refers to as omnipresent. God is one as omnipresent. Everything that changes is only that part of God that exists as weight. And weight exists inside the Universe that is moving around, and this moving around is happening as a place that is made of a cold, clear nothingness, which is the major portion that God exists as.

The next time you look at something as an object on the basis that God created everything, remember ***IT*** is not creating, ***IT*** is just reshaping. To say that something is created is to mean that something exists, and we see something as existing when it is matter, the product of the use and reshaping of ***ITS*** weight. God's creations are derivatives from that part of ***IT*** that is less than 1% of ***ITSELF***: The heated weight that ***IT*** existed as just before the Big Bang. After the Big Bang ***IT*** used ***ITS*** very dense form of weight to become the fragmented weight that ***IT*** now exists as, what we refer to as ***ITS*** creations. Scientists know these creations are pure energy transmuted--the same pure energy that existed before the Big Bang. So while we are still in the habit of using this word "create," ***IT*** is just reshaping ***ITS*** less than 1% weight.

I thank ***IT*** for redistributing ***ITS*** weight. Otherwise, I could not write about ***IT*** or what we understand as time and matter, or better

put, *matter* and time, since it is matter (weight) that changes and enables our human mechanical time system to exist.

And as always we have been focusing on ***ITS*** 1% weight, but since we now know that this weight is only that part of ***ITSELF*** that is less than 1%, I will from here on in my existence remember (until I get Alzheimer's) that God, ***IT***, also exists as nothingness.

As nothingness, ***IT*** is not reshaping or making anything. We cannot use nothing to make something. When ***IT*** did make something (a creation) ***IT*** could only do so by using weight combined with nothingness.

1- ***IT*** does not use this nothingness as nothingness to exist as matter, or at least the matter that we know exists as weight.

2- Matter can only exist by having this nothingness as the empty space between each atom's electrons and protons.

3- ***ITS*** nothingness does exist as a constant consciousness, for there has to be an operator for matter to know when it is supposed to transmute or move. ***IT*** has to be conscious of what ***IT*** is doing.

As for ***IT*** (God) being a constant, imagine what would happen if this constant way of ***ITS*** existence were not there; ***IT*** could develop Alzheimer's! And then who would run what now exists? I'd better get off this subject of ***IT*** developing Alzheimer's, for I may be the one to get it, for I would have forgotten to not stray from ***ITS*** nothingness.

Spirits

Have you ever given thought to our concept of spirits? I'd like to discuss this, but before I start, always remember that everything that exists is just really ***IT*** as one. So if spirits do exist, they too would still have to be ***IT***, and not merely millions of separate, independent spirits, for if there are any spirits, they would still have to be ***IT*** as omnipresent. I think that we, as humans, can have no knowledge of the existence of spirits because for spirits to exist in terms of something that we can percieve, they would have to be beings made of something, specifically made of matter (***ITS*** weight). Now, if spirits were material then they would not be

spirits, and if they happened to be made of a kind of nothingness, then again they would still be ***IT*** as ***ITS*** nothingness.

⌘~~~~~~⌘

The only one that is universal is IT, as ITS nothingness. **

❀~~~~~~~~~~~~~~~~~~~~~~~~~~~~~~~~~❀

IT as empty space

Empty space, or nothingness, is pure energy, ***IT***. No other name association is suitable to describe ***IT*** the way ***IT*** exists. ***IT*** is beyond human comprehension. ***IT*** is everything we see as matter and everything we do not see as empty space. If ***IT*** did not exist, neither could the illusion of the Universe exist.

A word for IT as nothingness

The words humans have been using throughout history, such as God, Creator, or Pure Energy to describe this block of nothingness and 1% weight that is our Universe do not bring us closer to the way ***IT*** exists and reshapes. Words as labels tend to denote that which is named as being something, which contradicts that which is nothing. Because of our travels in outer space, we have become used to thinking of empty space as something that exists; the human mind has accepted this even if we have not accepted that this area is also ***IT***. Our mind sees this area as being independent of God when this nothingness is evidence of ***ITS*** omnipresence. The mind does however scientifically accept this freezing cold nothingness as "pure energy." Still, I throw this question to the professionals: Is there a word for something (***IT***) that exists as nothing besides simply "nothingness"?

My feeling is that there might already exist a term for ***IT*** as nothing in mathematics. However, in the end, it does not matter what word we find because to understand ***IT*** better, you must look within yourself. As the saying goes, the kingdom of heaven resides within you. At our present stage of understanding, heaven has **become** lost between us and the Universe. If we look for ***IT*** within us, there is less of a chance of us getting lost.

The same question applies to the atom, such as the single hydrogen atom, that exists as 95% empty nothingness. What word we can use to denote the huge nothingness that exists inside every atom?

But I digress… The atom ultimately fails as the perfect model of our Universe's nothingness for the atom only came into existence after ***IT*** reshaped into the Universe we now have. A better depiction is a body as a shell or block that is made of what our minds can understand as being empty, filled with something that exists as nothingness.

Imagine this block as a 100% total: 99% is nothingness and 1% is the weight of the total block. I refer to this 1% as weight because protons and neutrons have appreciable mass, and once you have mass you have weight. Electrons, however, have very little weight and everything else is nothingness, which has no weight. In essence, if we remove the electron, the only weight left is from the protons and neutrons. What word might one use for this block that as basic elements is only 1% a block? Moreover, the 1% block is made of a transparent form of speed, and here too, this speed moves ***ITS*** weight as fragments that reflect light which we see as colors. And at the subatomic level, again, what is moving and changing is ***ITS*** weight, not the nothingness. I know this sounds paradoxical, but the truth is that something that is made of nothingness has nothing as something to use so that ***IT*** can effect a change.

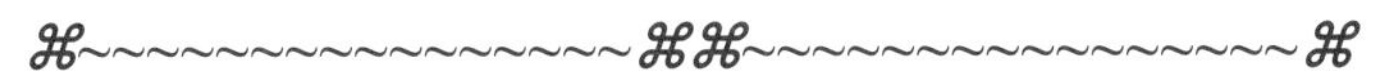

A hurricane is a phenomenon that has ITS weight on the outside and ITS emptiness on the inside and IT brings, at the immediate moment, seawater to the higher parts on dry land, such as mountains.

❁~~~~~~~~~~~~~~~~❁❁~~~~~~~~~~~~~~~~❁

Try it this way: nothing can change if it is made of nothing, so whatever ***IT*** is, this nothingness is how ***IT*** exists as a constant. ***ITS*** heated weight changes within ***ITSELF*** as the freezing cold nothingness that ***IT*** exists as and as the Universe that we know exists.

The emptiness that exists in every atom is just a continuation of the nothingness that exists outside, or just beyond the area where the electrons are. The difference between the nothingness that exists in every atom and Dark Matter is that in atoms their nothingness temporally separated by ***IT*** within ***ITSELF*** as high speed

(electrons), and this gives us the illusion of Dark Matter being something else, aside from the atoms' heated weight.

One nothingness

All the nothingness that exists, from the nothingness we call Dark Matter to the nothingness that exists in every atom is one nothingness as a whole.

Nothingness as energy

Science, and here I would like to thank the scientists that have made the present information available, makes it easier for my mind to accept that nothingness does exist. Otherwise, my mind would tell me that this is all madness. Instead, I can accept that God exists in omnipresence as a place that is scientifically known to exist as the nothingness of the Universe. Additionally, everything around me including myself is made of matter, which is also scientifically accepted to be at least 95% empty nothingness in the form of atoms within this nothingness. I must add though, that science has not been able to explain the nothingness or its purpose while many masters, as gurus, have tried to teach us about the nothingness of the Universe and how to connect to it in meditation. I would be lying if I said that I could do so for extended periods of time. But let me share with you what is known of the effect on people who do connect with this inner nothingness: They are at peace with themselves and with those around them. They are happier with themselves. What's more, ***IT ITSELF*** has always been leading us to and through ***ITS*** nothingness.

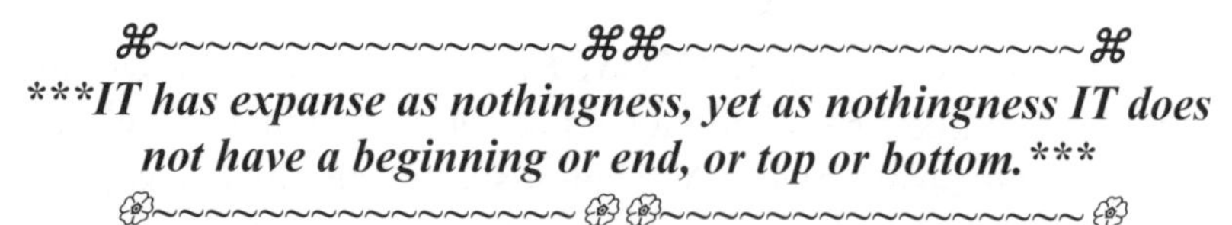

******IT has expanse as nothingness, yet as nothingness IT does not have a beginning or end, or top or bottom.******

Before we think that nothingness has no energy, let us consider that as ***IT*** exists, ***IT*** has energy as speed, which ***IT*** uses for moving light. We know that at 186,000 mps, speed carries the weight of a particle/wave from the Sun in a light beam. I have had to adjust my thinking to understand this duality of ***IT***.

When a light ray hits our bodies we feel the heat as energy that existed in the Sun, but colder. I say colder because the closer we get to the Sun, the hotter it gets; so the heat that left the Sun was hotter and got colder as it traveled through ***ITS*** cold nothingness. For this we should be grateful.

⌘~~~~~~~~~~~~~~⌘~~~~~~~~~~~~~~⌘~~~~~~~~~~~~~⌘

*** ***In order to understand ITS uncompromised nothingness, we cannot exist as ITS total nothingness, lest we give up ITS heated-weight as our human body.*** ***

⌘~~~~~~~~~~~~~~~~~~~~~~~~~~~~~~~~~~~⌘

The closer we are to the Sun, the hotter it will feel in temperature, but the speed of 186,000 mps remains the same, for this speed is a carrier. This could mean that ***MAXX-SPEED,*** (a speed faster than the speed of light which is what I believe to be the speed of ***ITS*** nothingness, of which you will hear more later on), is independent of temperature. I'm tempted to suggest that the total amount of energy that exists as heat could equal the total amount of energy that exists as cold, clear, high speed nothingness, but I will not, because in observing the way ***IT*** operates, one can see that it does not always do things in a uniform way. We can see this, for instance, in the atom where an electron has 1/1863 the weight of a proton. ***IT*** gets better results when ***IT*** is off balance or asymmetrical; which gives ***IT*** more reshaping possibilities. Still, the electron does have equal force as energy. And I have to accept that as ***MAXX-SPEED*** does exist as nothingness, therein also exists energy because energy is an intricate part of ***ITSELF*** as what is known as pure energy (God or ***IT***).

Looking at the duality of ***MAXX-SPEED***, which is also the duality of matter, I notice that, as matter, ***IT*** is more stationary, such as in the case of physical objects. We can visually see objects as having beginnings and ends, and we can see their locations. The opposite of this matter or weight, is something that cannot be seen and is not physically stationary. These are the qualities that ***MAXX-SPEED*** has. ***MAXX-SPEED*** cannot exist as weighted matter; it can only exist as a speeded nothingness, obviously, because the moment that this ***MAXX-SPEED*** carries ***ITS*** weight, it becomes light, or an electron. And here I would like you to remember that this high

speed that exists as light is moving outwards, which is the reverse of ***ITS*** weight, which is more stationary.

All energy moves something. ***IT*** as energy is moving ***ITS*** weight, which cannot be moved at ***MAXX-SPEED*** because the attached weight slows it down. For something to move faster, we therefore need more energy or energy attached to it as an object, at least in the world we have come to understand.

⌘~~~~~~~~~~~~~~⌘~~~~~~~~~~~~~~⌘~~~~~~~~~~~~~⌘

**** We can never exist as ITS total nothingness, without giving up ITS heated weight as our human body for we do have to use ITS weight in order to understand IT as ITS never changing nothingness ****

⌘~~~~~~~~~~~~~~~~~~~~~~~~~~~~~~~~~~~~~~~⌘

ITS mysterious ways

We as humans have always felt that ***IT*** works in mysterious ways. ***IT*** fascinates us the way a magician does. We know what we're seeing is an illusion; yet we are convinced all is real because of the way the magician puts his props together. Once we understand some of ***ITS*** props and workings, we may continue our normal way of living, but with a heightened understanding of what we perceive and of what put us and everything that exists in this Universe together.

It is like the illusion of a magical trick--the empty space that exists outside as the Universe. Since everything exists as one, and everything that exists within this Universe is one, it is sensible that our planet would appear to exist in a dark substance that is actually clear or transparent.

Empty space, for example, isn't actually "empty", for empty, in the sense of "nothing", does not exist. Even as empty space, something is there; ***IT*** is there as omnipresent as a way of existing.

And consider this: We are in outer space, deep space, a place that you think you have never been, yet we have been here always and without even being aware of it. I will use mathematics to explain since figures don't lie (and lies don't figure).

The whole Universe is 95% empty space. The Universe consists of less than 5% matter. Matter is composed of atoms that are 95% empty space.

It has been said that there could be billions of galaxies within our Universe. But for now, let us examine our galaxy known as the Milky Way. The Milky Way has approximately five billion stars within it; our planet orbits around one of these stars. When we refer to the amount of matter that exists within our galaxy, we first have to divide this 5% that exists as matter by the billions of galaxies that exist within the total 5%. (And this is an estimate that assumes matter is equally divided among galaxies: x % matter in our galaxy = 5% matter divided by billions of galaxies). Dividing the 5% of matter that exists in the Universe by billions of galaxies, we end up with a negative percentage, something to the minus of the billions. So now that we have established that the Milky Way, as a total, is somewhere in the minus of the billions, let's consider our galaxy, which has billions of stars within it, and determine its percentage of the matter in the whole Universe. Simply put, our Milky Way has billions of stars, not to mention planets and asteroids. These celestial objects comprise a minus to the billionth of the total 5% matter that exists within the 95% empty space.

A dark, cold, negative nothingness

We are referring to ***IT*** as negative because of the name we have given ***ITS*** weight as protons. If we add the words that we now also use to refer to the part of ***IT*** that is Dark Matter, the heading above would change to this: God, ***IT***, pure energy, basically exists as a dark, cold, negative nothingness as ***ITS*** invisible shell.

I add these few words because we have been trained by way of our education to see and understand that light is supposed to be positive, and darkness negative. I know that scientific minds and more educated people will adjust their way of seeing nothingness, including darkness, from being negative, and see ***IT*** instead for whatever it may truly be. We cannot change the way ***IT*** exists.

If ***IT*** is mostly made of a shell that is 99.99% cold, negative nothingness, and with ***ITS*** less than 1% weight ***IT*** permits us to exist and understand what ***IT*** can do with this less than 1% weight,

it is important to realize that we exist on this tiny amount of ***ITS*** positive weight.

If ***IT*** is more of what we understand as a nothingness, as the nothingness that ***IT*** has as a shell (what we call outer space), and as the nothingness that we exist as, due to the atoms that ***IT*** reshaped into, then we can see and understand, at least intellectually, that something, ***IT***, so great and powerful, does exist, and ***IT*** is continuing to educate us more about what ***IT*** could be like, independent of the way we use words to understand ***IT***. I know that when I go inside (that is, when I meditate) to be with ***IT***, I cannot use any of the habits or language I express outside my body. I have accepted the way ***IT*** exists within me. I am again thanking you (***IT***) for permitting me to mention your ***(ITS)*** existence.

ITS dual force

Now I'd like to address the subject of ***ITS*** nothingness as a force that could be equal to ***ITS*** weight: I feel that it is not in the nothingness, but rather in the coldness that exists as this vast nothingness where we will find the opposite force to ***ITS*** concentrated heat.

It makes sense to see that ***ITS*** less than 1% heat could be an equal force to ***ITS*** 99.99% coldness, for what we are seeing and touching as matter, or as something that is made of something, is because of this less than 1% that ***IT*** is as heat. Just as the opposite of something is nothing, (as in ***ITS*** 99.99% nothingness), the opposite of heat is cold.

As ***ITS*** nothingness, ***ITS*** force may be in the coldness, and as heat, ***IT***, as a force, is pushing outwards, which we can see as the outward push caused by the Big Bang. With the aid of ***ITS*** cold-pulling nothingness taking place within ***ITS*** cold invisible shell, we have the opposite of heat; that is, cold, which exists as ***ITS*** shell that is pulling inwards to hold ***ITS*** shell together.

Since ***IT*** exists as one, and ***IT*** operates as two in two of ***ITS*** stages (stages 1 & 2), ***IT*** exists as opposite forces. ***IT*** uses ***ITS*** heated weight in an outward pushing effect as a way to push ***ITS*** weight,

but ***IT*** is also held in by ***ITS*** cold pulling force because ***IT*** has to stay within ***ITS*** oneness, for this heat cannot exist outside of ***ITSELF***, not even as an independent heat force.

As discussed in ***ITS*** three scenarios, as ***ITS*** third stage, ***IT*** exists in ***ITS*** most balanced state where ***IT*** is not pulling or pushing as opposite forces. At this stage, there is no reshaping taking place; ***IT*** exists as a clear nothingness that has no concentrated heat or coldness. And as always, whatever the truth of the matter may be, it will only help us in understanding ***IT*** better.

⌘~~~~~~~~~~~~~~~~~~~~~~~~~~~~~⌘

*** ***Antimatter is ITS nothingness*** ***

❀~~~~~~~~~~~~~~~~~~~~~~~~~~❀

Atoms may be 99% nothingness

IT does things in a repetitive manner; therefore, I will venture to say that instead of the atom being 95% empty nothingness, it may be closer to 99% empty nothingness. Most scientists would agree that matter comprises 4.5% of the Universe. If we apply to this the 95 plus % emptiness that exists in the atoms that make up this 4.5% matter, the result would compare to the 99% nothingness that ***IT*** exists as the Universe.

No theory, concept, or philosophy claims that the Universe is 95% emptiness. Regardless, ***IT*** is there, even if we are only beginning to explore and find more information on this 95% cold, clear, empty nothingness that undoubtedly exists. Furthermore, I am sure that ***IT*** will show us how to open the door to the explanation of this part of ***ITSELF,*** for ***IT*** is the greatest teacher that we have ever had, which has always existed and will always exist. We are not in the Dark Ages anymore. We are now in a very advanced stage of thinking, and we have the use of ***IT*** reshaped into the latest technology, plus what is yet to come.

So, if ***ITS*** composition is more in the area of being 99% nothingness, then this may also be true for atoms, for they too are made in ***ITS*** own image, as a way of being. However, we will have to hold off on this until we can find a surveyor that can calculate more exactly how much empty space exists inside the atom. I am sure that whatever the exact number is, it will not change the way

IT is or what we produce from atoms.

⌘~~~~~~⌘

Help wanted: A calculator that can calculate ITS nothingness.

❀~~~~~~~~~~~~~~~~❀❀~~~~~~~~~~~~~~~~❀

IT as one nothingness.

IT exists as one total nothingness, as ***ITS*** shell, which behaves as an inner and outer shell simultaneously, and ***ITS*** weight is inside the shell. Life is a part of ***IT*** where ***IT*** has taken ***ITS*** fragmented weight that exists within ***ITS*** total nothingness and has given this fragmented weight mobility and an appearance in order to exist as something, by placing a tiny amount of ***ITS*** weight in the form of electrons, protons, and neutrons, so that ***ITS*** weight could be visually seen, as matter, so that when we see ourselves as something that is alive, what we are really seeing is ***ITSELF***. If I start with what I see as something, let's say a living person, what really exists is this: first, the person is there because ***IT*** took a tiny amount of ***ITS*** weight and attached ***IT*** to ***ITS MAXX-SPEED*** (which you will read more about later on in this section) so as to become electrons. ***IT*** then took another, larger, quantifiable portion of ***ITS*** weight, and surrounded this bigger weight which we call protons and neutrons, with what ***IT*** had previously reshaped into as electrons, so that what we call matter could exist. In between these, the protons and neutrons on one hand, and the electrons on the other, is the nothingness, the 95% empty space that exists inside the atom.

Now, if we go back to discussing the person that is alive again, what is happening is that the reason why we can see a person as being physically there is due to the fact that ***IT*** took a very tiny amount of ***ITSELF*** as ***ITS*** weight to become the electrons that exist as the person's outermost layer, which is alive, and then ***IT*** formed ***ITS*** other fragmented weight into the protons and neutrons that make up the atoms that compose the person's body, so that when we see this person, we see him/her because of the way ***IT*** is using ***ITS*** weight(matter).

Now let me return to ***ITS*** nothingness again and go back to the concept of illusion, to just what it is that is driving these electrons

with ***ITS*** way of existing as a high speed (***MAXX-SPEED***) around ***ITS*** other weight as protons and neutrons. Now here is what is very interesting: ***IT*** is doing this within ***ITS*** total nothingness, so that when you see a person, what has really occurred is that ***IT*** took that very small amount of ***ITS*** weight as electrons, gave ***IT*** distance or separated ***IT*** apart from that other part of ***ITSELF*** that exists as ***ITS*** fragmented weight as protons and neutrons in order to form atoms, and gave ***IT*** what ***IT*** already exists as. Here I am not speaking of mere life, but of the divine consciousness which is ***ITS*** being, as the nothingness that is always a constant.

So, returning to us being alive, this is the condition in which ***IT*** gave ***ITS*** once total weight (the weight that existed at the moment of the Big Bang) and fragmented this weight into electrons, protons, and neutrons, so that ***IT*** could reshape into the matter that we exist as. In addition, ***IT*** gave this weight certain functions, like our brain, heart, liver, lungs, and certain properties so that this weight could have mobility, so that ***IT*** could exist as ***ITSELF***, as a divine consciousness, which we call being alive.

Again, when we see humans, each one of them and everything else exists within ***ITS*** one total nothingness, so that even if we get the impression that we are outside of something, we are really still inside of ***ITS*** total nothingness, and the reason is that ***ITS*** nothingness exists in everything that exists, such as air, water, trees, houses, trains, planes, and every other thing on and in this planet; all is inside of ***ITS*** total nothingness, because anything that does exist, in order to exist, has to be made of ***ITS*** weight which has to exist within ***ITS*** one total body that exists as what we see as this Universe.

Returning then to what I started out to say, ***IT*** is very interesting to see things as being out there, in what we call reality. Nevertheless, everything out there is really inside of ***ITS*** one total nothingness.

All of this will make more sense if you always remember that nothing is really ever created or destroyed as pure energy, for everything that has ever happened or will happen, has always been and will always be ***IT*** as one.

You might understand ***IT*** better this way: It is the separation of ***ITS*** fragmented weight that gives us distance (or extension in space) within ***ITS*** one total nothingness, which gives us the illusion of there being trillions or googols of things existing out there, when in reality, there is just one of ***IT***, where all ***ITS*** fragmented weight is moving within ***ITSELF***.

The Universe as the size of an atom

Information is now available that points to the scenario that the whole Universe was once the size of an atom as dense weight. This is what we believe was happening when ***IT*** was in stage one, when ***IT*** had all its weight in one place as the very dense matter that existed before the Big Bang. But this matter that may have been the size of an atom still existed within ***ITS*** invisible shell that still exists as this Universe. So ***ITS*** dense weight may have been the size of an atom but not ***ITS*** 99% nothingness.

I agree that ***ITS*** size, as heated weight, can be compressed, but I do not see how nothingness can be compressed. ***ITS*** weight can be measured because it can be seen by us as something that exists, and if we could know ***ITS*** actual size in terms of ***ITS*** weight, then we might have a better understanding of ***ITS*** actual size as ***ITS*** nothingness, in terms of extension, which is the place inside of which ***ITS*** weight exists. But ***ITS*** weight as dense matter does not reflect the cold, clear, nothingness that composes the rest of the Universe. We know and accept scientifically that this nothingness exists and always has, but ***ITS*** nothingness is independent of the size of ***ITS*** weight, which can never exist outside of ***ITSELF***.

And, ***ITS*** size makes no difference; ***ITS*** immensity is beyond our ability to measure. How can we measure ***ITS*** size if it is basically 99% nothingness that has no beginning or end? The only measurement we can take will be related to the distance between the fragments of weight ***IT*** exists as, in terms of celestial matter, within the 99% clear, cold, invisible nothingness that ***IT*** has as ***ITS*** body shell.

As I have mentioned before, we know that this Universe is pure energy, and we have watched how ***IT***, as pure energy, behaves; but

we still have a hard time seeing the Moon as God, even when we know, scientifically, that the Moon, having been formed by the reshaping of this pure energy, is part of the totality that is God.

There may be a mathematical way of measuring ***ITS*** extension, which would have to be based on a way that we could measure the total area occupied by all ***ITS*** weight. The only other way would have been to measure the nothingness before the Big Bang. This nothingness has occupied the same extension in all three of ***ITS*** stages of existence. So that if we see that this hugeness has always existed as ***ITSELF***, as a cold, clear constant, invisible nothingness, and ***ITS*** weight has always existed within ***ITSELF***, this would be applicable to all of ***ITS*** three stages of existence during which only ***ITS*** weight has been changing within ***ITS*** cold, clear nothingness of a shell, which is what this Universe is now in as ***ITS*** second stage.

Let us say that we measure ***ITS*** total weight, and ***ITS*** total area is one mile. This would mean that ***ITS*** size, as cold, clear nothingness, would be close to 99 miles across. When I stop to think of ***ITS*** size, using the information that exists, such as the size of our galaxy, the Milky Way, which we have information on, our galaxy's size as ***ITS*** nothingness would be on the basis of ***ITS*** 99% nothingness. This would mean that if we could compress just this area that we call the Milky Way, so as to remove all of the nothingness, the compressed matter could be smaller than the size of Earth! Nevertheless, I see nothing wrong with whatever size we may think ***IT*** really is, for this will not affect the way ***IT*** exists or operates.

For now, with our limited knowledge of ***ITS*** real size, all we have to do is be careful not to bump into ***ITS*** other fragmented forms of weight that exist within ***ITS*** cold, clear nothingness. We should be grateful that ***IT*** exists as a clear nothingness; otherwise we would be having a lot of accidents crashing into the fragmented weight that ***IT*** exists as, such as the celestial bodies that are the inner self of ***IT***. We should be also grateful that, as our future astronauts travel around ***ITS*** invisible shell as a body, they will never really

get lost in what we call outer space, for they will always remain inside of ***ITS*** cold, clear, conscious, invisible nothingness.

And if we look at the way ***IT*** was at stage one, we will be able to see that just before the Big Bang ***IT*** had all its weight within ***ITS*** cold, invisible body shell that exists as a form of nothingness. From our knowledge about this stage of ***ITS*** existence, we know scientifically and mathematically that ***ITS*** weight can become infinitely smaller, to the point where ***ITS*** weight would just disappear. When this happens ***IT*** will then be in stage three. Let's return to the atom to see this situation better.

In the hydrogen atom, for example, ***IT*** has less than 1% as weight occupying space. When the electron is removed, ***ITS*** weight, as the proton, will decay, becoming smaller as the proton releases ***ITS*** weight and turns into energy and even smaller particles and waves that are carrying ***ITS*** weight elsewhere as ***IT*** continues to reshape.

Now that we know more about how ***ITS*** weight can become smaller than a proton--the weight that enables the atom to exist and also that enables us to see things as existing--we can see how ***IT*** could have ***ITS*** weight reduced to something that could still exist as weight, yet be so small that we would not be able to see it. So that if all the now existing weight within ***ITS*** 99% nothingness were reduced to an infinitely smaller weight and was redistributed throughout ***ITS*** 99% nothingness, ***IT*** would then be back to what I call ***ITS*** stage three, where ***IT*** could bring in all ***ITS*** weight into a central, singular point and back to the instant before the Big Bang (stage two). We should remember that if this is so, it may not be the first time that ***IT*** has formed a Universe such as the one we have now.

We should be grateful that ***IT*** is always reshaping into something new with ***ITS*** weight, which exists within ***ITS*** clear, cold nothingness, which is that part of ***ITSELF*** that is a living, divine consciousness. After all, if ***IT*** now exists as a living, divine consciousness as ***IT*** reshaped into us, then ***IT*** exists this way as ***ITSELF***. I can assure you that I did not give my children this living, divine consciousness that they exist as. And I am now aware that neither did my parents give ***ITS*** divine consciousness to

me. We are all a gift that ***IT*** has given us as a chance to exist as ***IT***, as a living, existing divineness that we are now hopefully enjoying, as the life that we are living. If you are not enjoying life, then my message to you is to ask ***IT*** for help and guidance, for as you can see, ***IT*** is beyond any human master at doing things.

For now, I will enjoy ***ITS*** existence as myself and as those around me, for even if I did know how big ***IT*** really is, ***IT*** is not going to give me a better piece of candy than what ***IT*** already has. ***IT*** is every atom that you are using to read, think, and exist.

Gravity and nothingness

Gravity is how ***ITS*** weight is pulling to bring together ***ITS*** fragmented weight. The opposite of this is the movement of ***ITS*** nothingness, which stemming from the Big Bang theory, is pushing its weight away. So if I look at ***ITS*** weight (atoms or matter), which is what makes things visually possible, or as something palpably existent, and then I take ***ITS*** reverse, which is ***ITS*** nothingness, I begin to see that there is no way we will ever be able to see this area in which ***IT*** exists as a nothingness. And I now also understand this phrase that we refer to as "pure energy has no beginning or end," for this part of ***ITSELF*** is made up of a cold, clear, high speed nothingness, and this speed itself makes it virtually impossible for us to see.

Imagine us trying to see something that is clear and is moving faster than 186,000 mps! Since we are used to seeing things (***ITS*** weight) in order for things to exist, it will be impossible; we will never be able to see ***ITS*** 99.99% cold clear nothingness. We will have to accept this on the basis of faith; ***IT*** exists as this 99.99%, for we know that this area known as the Universe does exist as a cold, empty nothingness; otherwise, we could not exist within ***IT***.

I only hope that ***IT*** will let me stay in this existing moment so I can see what other information we discover related to ***IT*** as ***ITS*** nothingness. There, in fact, should already be a lot of information stored in people's minds and in computers that when reviewed will shed new understanding on ***IT*** as this cold, clear, speeded nothingness that ***IT*** also exists as.

Now that we are aware that by looking for ***ITS*** opposite, as weight, on which we already have a great deal of information, we will be able to discover more about ***IT*** as cold, clear nothingness. We also already have a great deal of information on ***ITS*** nothingness as light and electrons, which reminds me of the phrase "there is always something new under the Sun," which is still true, for the Sun is light, and light is speed, as something that exists as a nothingness. So, I say to all: Full speed ahead on gathering more information on ***ITS*** clear, freezing cold, speeded nothingness!

IT has girth

ITS clear nothingness exists simultaneously as an inner and outer body (structure, existence, being). ***ITS*** weight exists within ***ITSELF*** as ***ITS*** body, and ***ITS*** inner nothingness wraps ***ITSELF*** within ***ITS*** inside as one, for there could not be an outside. Where would this outside exist as omnipresent as ***ITSELF***?

As you can see, it is difficult to describe what we know is there but which lacks a form that we can see or touch. ***IT*** does not have a border that we could use as a reference point from which we could begin to measure ***IT***. However, maybe mathematically we can find a way to calculate ***ITS*** size because mathematicians have ways of knowing nothing as something. But for now, ***ITS*** girth is so huge, that just as in the case of the span occupied by our galaxy, I, for one, will not wander off. I can instead feel ***ITS*** presence within me as ***ITS*** constant nothingness in meditation and as ***ITS*** weight as my human body that has a constant tendency to increase in weight!

⌘~~~~~~⌘

****** The Universe is not expanding. ITS weight is moving outwards within ITS nothingness . ******

~~~~~~~~~~~~~~~~~~~~~~~~~~~~~~~~~~~~

One thing that I have observed is that ***IT*** always moves ***ITS*** weight, as celestial bodies within ***ITS*** nothingness, and even when ***IT*** fragments ***ITS*** weight as atoms, ***IT*** still keeps ***ITS*** weight as protons, neutrons, and electrons within ***ITS*** total nothingness and as the Universe and as the matter that exists within the Universe as ***ITS*** nothingness.
~~~~~~~~~~~~~~~~~~~~~~~~~~~~~~~~~~~~

Furthermore, ***IT*** has a sense of this size. Many animals can detect distances as a security measure. And humans can sense distances too, so ***IT*** might also know ***ITS*** size, for ***IT*** incorporated the sense of distance in us. And, ***IT*** just might at some moment let us know how big ***IT*** is in extension. Although even then, we might not understand ***IT***, for it is hard to imagine that our whole planet is not even the size of a grain of sand when it relates to the hugeness of ***ITS*** existence.

This invisible nothingness that ***IT*** exists as does incorporate distance, as the distance that exists between the fragments of matter inside ***ITS*** cold, clear nothingness that exists as this Universe. We know in fact that at the very minimum, ***IT*** spans 250 million light years away in distance as scientists have found signs of a black hole that far away, although naturally these 250 million light years were also determined by the human mind. [2]

I have to remind the reader that when we refer to distance, we do so because we are using a measuring system that works well for us but is not universal. Some people, such as those who have seen Star Trek, are not aware that our interplanetary friends do not measure in inches, meters, or miles. My only assumption as to why films depict this is because English has been established as the most widely-used language for communication in international travel, so it has become the same for interplanetary travel.

That our minds can ask or understand anything at all is only because ***IT*** has reshaped from where ***IT*** existed as our primitive beginning to where ***IT***, as our civilization, now exists. I am sure that we will understand ***IT*** even better as we are permitted to see ***IT*** as the technological society that we, as ***IT,*** are reshaping into.

But getting back to ***ITS*** size, this will become a problem because, as I have mentioned before, the only distance that we can measure is the distance that exists from one body of matter to another, such as between planets and stars and other celestial bodies. And, we

---

[2] Nowadays scientists believe the Universe is at least 10 billion light-years in diameter.

use the same measuring system established for our convenience here on Earth.

If we try to measure ***ITS*** size as if ***IT*** had an outside, first we do not have anything else to compare ***ITS*** size to, and second and most importantly, we do not know how to measure something that is made of nothingness. ***IT*** does not have a point as a beginning or as an ending from where we can start measuring. For example, one might describe ***ITS*** nothingness as being housed in a shell, but this word isn't adequate; it implies we can visualize ***IT*** having a hard exterior like that of an egg, which would be made of matter, which would have weight. It makes sense that we would imagine this as we have only been able to measure ***ITS*** weight as the matter that exists within ***ITS*** nothingness as this Universe.

So, for our observation of ***ITS*** size using the weight, as matter, that exists within ***ITS*** nothingness, we would only be able to see ***ITS*** weight inside ***ITS*** clear, cold nothingness. We would not be able to see ***ITS*** cold, clear nothingness as having a beginning or ending to which we could try and measure ***ITS*** distance from one side to the other, if indeed ***IT*** does have "sides", or better yet, from ***ITSELF*** to ***ITSELF***. All we can do is continue measuring the distance from fragment of weight to fragment of weight within ***ITS*** nothingness.

If there is someone who knows of a better word for ***IT*** as ***ITS*** housing as a nothingness, please contact me.

IT has no outside

Whatever ***IT*** is, ***IT*** does exist, and in ***ITS*** existence, everything that is happening as ***IT*** is happening within ***ITSELF***, as what is now happening within this existing Universe, as ***IT*** as omnipresent, or as what some understand as God or the pure energy that this Universe exists as. Knowing this you will understand why ***IT*** does not have an outside, because, naturally, anything that could exist outside of ***IT*** would not be part of ***ITSELF***.

At least we know that anything that has happened or will happen as ***IT*** reshapes, has to happen within the cold, clear nothingness that

IT exists as. If ***IT*** were now existing as any of the three previously described scenarios, not having an outside would still apply to ***IT***.

And if I could see ***IT*** in ***ITS*** third scenario, which is when all ***ITS*** weight was evenly distributed within ***ITSELF***, then ***IT*** would be even harder to see ***IT*** as having an outside, for in scenario three, ***IT*** would exist as a clear environment having an even temperature. All we would see of this stage is nothing, similar to what happens when we see things through clear air, where we are not even aware of the air being in front of our noses. And I can definitely tell you that I do not know what would be happening within ***ITSELF*** as that stage. The same way we do not know what was happening within ***ITSELF*** as stage one, which was when ***IT*** existed before the Big Bang, when, if it were possible for us to see this stage, all we would see is again nothing, for all ***ITS*** clear weight would be found in one concentrated spot, or place. Although, we would know that ***ITS*** weight would be there because of the concentrated heat that would exist as that singularity.

I would like to recall some previous information that may help you to understand why we would not be able to see this concentrated, clear, heated spot. It would be like seeing things outside yourself that are there because they are made of substances that have colors. Between you and the colored objects that you are seeing, there exist millions, billions, and trillions of atoms that are transparent because they have no mineral coloring to them. You can see right through ***ITS*** clearness, that also exists as ***ITS*** weight as the protons and neutrons that are in all those heated atoms that are there as air. We cannot see them, but they do exist as part of ***IT.***

So again let us be grateful that ***IT*** does exist, even in this clear form, otherwise it would be harder for us to read or see our televisions. If ***IT*** did not exist as this clearness, then there would always be something to tint our view of what is outside of us, and this would most likely give us an excuse for the accidents that we are having with our motor vehicles. And this tinted view would also cause a problem to what we would see as ***IT***, as what is out there as ***ITS*** invisible shell called the Universe.

Nothingness as the Universe

Imagine opening a door, and through it you can look into the Universe. You see that there is nothing to your left or right or up or down. The reason you cannot see ***ITS*** borders is because ***IT*** is made of a freezing cold nothingness. You see the planets and other matter in our solar system because of the small amount of light that ***IT*** has reshaped into as stars.

In this clear empty space, due to ***ITS*** nothingness, you can see things moving about; you see these objects because ***IT*** has fragmented ITS total weight in to smaller chunks of ***ITSELF.*** And the reason why you can see ***IT*** as ***IT*** moves is because ***IT*** slows down from 186;000 mps where ***IT*** cannot exist as weight to where something can exist in the form of matter, so that ***IT*** can continue to reshape into infinite objects or situations.

Divine Consciousness

While I am in no way an authority on this subject but only an observer, I am going to venture into the meaning of the word consciousness; the consciousness that we are. As with some of the lower animals that came before us, we too are the result of this consciousness that has been developing in living creatures.

We have a three pound brain which is 85% water. With it we are able to think, which is necessary for our primary programming: survival. This is what ***IT*** is mostly concerned with so that the reshaping can continue.

We have reached our present development with this new hi-tech society we live in. Because of it we now are capable of understanding what would have been impossible years ago. ***IT*** has made it possible for us to discuss this subject of consciousness. So, in order to be aware of consciousness, I have to know that I exist as all these atoms that energize me to write this. But I have to be conscious that I am also the same nothingness that ***IT*** is. To even think about this I have to know that I am a part of the totality that ***IT*** is.

Also, to understand the meaning of consciousness I have to remember that I did not put it there. Consciousness is ***IT*** as life

with mobility; it is what ***IT*** allows. The consciousness of lower animals is the first programming by which they survive; the second programming is the one to reproduce. I look for this consciousness inside of me and feel that it should be somewhere inside the mind, inside my brain. Yet, I know that ninety five percent of me is nothingness. If my mind, as consciousness, is not made of matter; my consciousness must be a part of ***THAT*** nothingness.

When I connect to ***IT***, I do so in meditation. Using the third eye, I am connecting to a form of nothingness that exists within me in a given area. The biggest problem I have connecting to ***IT*** is detaching the mind from all the thinking, whispering, and chattering that goes on inside me.

Finding that inner place where nothing controls me, where no wandering thoughts disconnect me, and realizing that there is no danger there, is a very conscious experience. The mind can be part of this peaceful nothingness.

Since I was a kid, I could see my environment, people, streets, and places well enough to survive. Through my eyes, my mind learns what it has to take control of, for who are we if not what we think we are? But as a child I didn't know that it is natural to think in terms of things; it is what the mind has to go on; it thinks, for it to exist. In this way, we can say that we are connected 100% to matter from the beginning of our existence as matter, things, places, and people. And I am glad to have learned this from many people, including Maharaji, who exists to show us how to connect with the inner nothingness.

Many names are used to describe this inner nothingness including the "inner self," the third eye, or simply a peaceful place. For that I am grateful to ***IT*** who never forgets to send someone to help us connect with that part of our divine consciousness--a consciousness existing as nothingness. And we cannot control this because our minds are made to deal with matter unless the mind refocuses and accepts that in nothingness is how ***IT*** exists, and does so as 99.99% of everything.

Remember that ***IT*** is not being created or destroyed. Forget about yourself and think of ***IT*** as something that does exist and does things in ***ITS*** own image. The part that ***IT*** exists as nothingness is where ***IT*** exists as consciousness, and since consciousness is a form of nothingness, it does not participate in change; for change only happens to ***ITS*** weight. Consciousness is where the divine consciousness exists, and it is in this divine conscious nothingness where the majority of ***IT*** resides. The divine consciousness that ***IT*** exists as can be found mostly easily in this 99.99% nothingness.

At the same time, the weight of ***IT***, as the matter composed of the atoms that are also surrounded by nothingness, makes ***IT*** easier for ***IT*** to be conscious of everything outside us. ***IT*** is all knowing, but not as our minds, because the mind is only ***IT*** in the form of water based matter. Our challenge is to distinguish between what we see because of the mind and that of which we are conscious because of our existence as part of ***IT***.

If I did not have this consciousness, I could not understand myself and my environment. Being conscious is not because of the water and all the other materials that I need in order to function to live; it is the same divine consciousness that ***IT*** exists as. ***IT*** is everything that exists, and this divine nothingness (life) is something that permeates the entire Universe. And again I reach for a word to describe this something that exists as a body, house, or shell that, to our sense of reality, is made of nothingness.

I have tried to see all of this in different scenarios, and to aid your understanding, let us go back to just before the Big Bang where ***IT*** existed as a concentrated oneness, and again, I have to stop to remind the reader that oneness is ***IT*** as very dense matter and 99.99% nothingness. Before the Big Bang, had we been around, we would have been able to see ***IT*** in ***ITS*** 99.99% nothingness because atoms did not exist. So for us to say that at this pre Big Bang stage ***IT*** existed as oneness, we would have to include the 99.99% nothingness surrounding the less than one percent heated weight that exists as matter.

My feeling is that whatever ***IT*** is now, ***IT*** too was ***IT*** then, as a way of existing. If I am alive and conscious, then ***IT*** has always been alive and conscious. For me to see that ***IT*** reshaped from the form ***IT*** had before the Big Bang, then ***IT*** too had to be conscious of what ***IT*** was shaping into, which is everything that now exists.

And to top it off, the same way it is hard for us to understand that a God can be 99.99% nothing, it is also hard for us to see that since ***IT*** is everything that exists, as ***ITSELF***, everything that is alive is really ***IT*** as one. ***IT*** has always been ***IT***, even before the Big Bang, and it will still be ***IT*** after this Universe reshapes itself again, as it expands or contracts to become one more existing possibility.

In addition, it is not the Universe that will expand or contract, it is ***ITS*** weight that is giving us the illusion of the matter that will expand or contract. ***IT*** has always stayed as ***IT*** is: a constant, cold, conscious, empty nothingness.

The Third Eye as nothingness

The next time you meditate, connect to nothingness and then return, remember this: The third eye sees the nothingness that is ***IT*** as ***ITS*** own constant.

ITS nothingness as omnipresent

We know that the meaning of the word omnipresence is that ***IT*** is in all places at the same moment. There are two ways to understand that statement: first, that it is ***ITS*** nothingness that is in all places at the same moment as the nothingness that exists as this Universe, and also, as the nothingness that exists inside every atom, and second, as for ***ITS*** weight, being in all places at the same moment is a only possible as being inside of ***ITS*** one total nothingness as a place. Let me explain this: The fragments of weight (atoms) that our whole planet exists as are not the same fragments of weight that exist somewhere else inside of this Universe, so that all ***ITS*** fragmented weight is not in the same place (in the same extension of space) at the same moment, but ***ITS*** cold, clear nothingness is in all places at the same moment and let me add that, since ***ITS*** fragmented weight exists within ***ITSELF***, within an area that exists at what I call ***MAXX-SPEED***, the distance that exists in-between ***ITS*** fragmented weight (matter)

exists in an area where the distance from one fragment of ***ITS*** weight to another is covered by this ***MAXX-SPEED***.

Look at ***IT*** this way: Using our understanding of time, we know that at the speed of 186,000 miles per second time stops, in the sense of there being no time difference in getting from one place to another, or as in there being no individual fragments of matter. Or to rephrase that: It is only what we call time, that separates one place from another, and this perception that we, as finite beings, have of time, is what prevents us from knowing that this Universe as ONE, as when we speak of omnipresence.

This area that ***IT*** exists as is a nothingness where this ***MAXX-SPEED*** exists as being timeless. So, since ***ITS*** body or shell is made from a form of a speeded nothingness, this is why ***IT*** exists as being timeless.

Now, since this ***MAXX-SPEED*** is faster than where time stops, this whole area that exists as ***ITS*** nothingness is an area where our time system cannot be applied. Please follow me closely now: So, considering our planet, which is not large but nevertheless it still exists within this ***MAXX-SPEED*** area, we must say that anything that happens at any point on this planet still exists within ***ITS*** oneness as this ***MAXX-SPEED***. Another way to state this would be to say that the area which our planet occupies falls within this area of ***MAXX-SPEED***, which means that everything that is happening on our planet exists within an area where time does not exist on account of this ***MAXX-SPEED***. Now, if we take our next nearest planet, we have to remember that every centimeter that exists in between our planet and the next, exists in an area that is also timeless, for there really isn't anything there to apply time to, in terms of matter, because what does exist as the "in between" is really just ***IT*** as ***ITS*** omnipresent nothingness. Furthermore, the distance that exists between planets and other celestial bodies also exists within this area of ***MAXX-SPEED***, where time does not exist.

Looking at the same scenario from another angle, at the outermost layer of anything that exists as matter, you will find ***IT*** as ***ITS*** fast

speed. Let's take anything that exists as matter, which is made from atoms, and the outermost layer of any atom exists as ***ITSELF*** as a high speed nothingness. Now if you take one of these tiny atoms' circumference, the outermost area is where the electrons exist as a high speeded nothingness, and as you travel inwards, this area is followed by ***ITS*** nothingness, and then you find the atom's nucleus which is composed of tiny fragments of ***ITS*** weight in the form of protons and neutrons. These fragments of ***ITS*** weight are so tiny that they are practically invisible. For this reason you may understand better when I refer to all this just being an illusion, because everything that exists is made from this speeded nothingness in the form of electrons, followed by ***ITS*** empty nothingness, and then with a very tiny amount of ***ITS*** weight. It is only when ***IT*** packs billions and trillions of these invisible, empty, tiny atoms that we begin to perceive the things that exist as matter as being there.

Now remember that you and I, and everyone else, is made from these tiny invisible, empty atoms that ***IT*** reshaped into, as the pure energy that ***IT*** exists as. Another way to imagine this place that exists as omnipresence is to remember that everything that exists as this Universe, exists within this ***MAXX-SPEED*** nothingness, making everything in this Universe that exists as ***ITS*** weight exist in a place where this nothingness unites everything as one place. In order to understand this you must remember that ***IT*** exists in all places as the same moment as ***ITS*** nothingness, and since this nothingness is not made of something that our minds can comprehend, because this nothingness, does not have ***ITS*** weight for us too see, and because ***ITS*** huge shell body is composed of something that again our minds cannot comprehend because ***IT*** is composed of a speeded nothingness, which is one of ***ITS*** dual forms of energy, this is why when we view the Universe, we focus on what we can see, which is really ***ITS*** weight, and our minds bypass the emptiness that this Universe now exists as. One reason for this is because this emptiness is a clear form of energy that is composed of this ***MAXX-SPEED***, which we will never be able to see. However, our mind does accept that this empty Universe that definitely exists is there, and is so huge that we call ***IT*** infinity,

which is easy to understand because we are referring to ***ITS*** nothingness. Therefore, when we refer to this Universe as being infinite, we are referring to ***ITS*** cold, clear nothingness, and when we refer to the things that God can create as being infinite, we are referring to the infinite possibilities that ***IT*** (God) can reshape ***ITS*** weight into, within ***ITS*** nothingness. In conclusion, we can say that the one common thing that the word omnipresence refers to is ***ITS*** nothingness.

$E=mc^2$ plus nothingness

We have a lot of information on this equation, $E=mc^2$, as to how much energy matter has. In the equation, E = energy, m = mass, and c^2 means to multiply the mass by the speed of light squared (multiplied by itself, that is, 186,000 mps x 186,000 mps). This equation is of course only applicable after the Big Bang because it relates to the speed of light, and light only came to exist after the Big Bang in the form of stars. Additionally, to say energy equals mass is also only relevant after the Big Bang because what we now have as mass only comes after the formation of atoms having protons, neutrons, and electrons, ***ITS*** heated weight. Furthermore, heated weight can only exist when surrounded by nothingness, as in the empty space inside any atom.

E and m are therefore part of ***IT*** as the energy that ***IT*** exists as in ***ITS*** heated weight, for heated weight permits matter to exist as the mass of protons, neutrons, and electrons. The speed of light also exists as ***IT***, as is evident from the speed of electrons. Yet missing from the equation is ***ITS*** nothingness, as the empty space that exists in all the atoms.

The equation $E=mc^2$ permitted us the understanding that led to the atomic bomb where we could see what energy ***IT*** is composed of, for this energy came from ***ITS*** heated weight, a small fraction of ***ITS*** total weight. The atomic bomb produced heat and light. The heat came from the weight of protons and neutrons. And light, remember, is a by-product of the speeded electrons. The atomic bomb released heat as the positive, and speed as the negative, colliding with each other to produce an emerging light. The Big Bang as an explosion did the same thing; it brought together ***ITS***

heated weight, as a positive, and ***ITS*** speed as a negative, which exploded into stars and all the matter of the existing Universe, which again brings me back to ***ITS*** nothingness, because if I look at the atomic bomb, the heat and speed were transmuted, and wherever this transmuted energy went, so did the nothingness. Maybe as nothingness, ***IT*** is not doing anything other than existing as ***ITS*** constant consciousness.

To release the energy that ***ITS*** weight as mass has, I have to include ***ITS*** nothingness as the speed of light. To do this, I must remember that the speed of light is something that we are mentally using to understand ***ITS*** speeded nothingness. Before the Big Bang, there were no stars, but there still existed ***ITS*** weight as mass, and there still existed ***ITS*** high speeded nothingness (***MAXX-SPEED***). Even the speed of light, as nothingness, can only exist in a vacuum where none of ***ITS*** weight is present, because as soon as light hits another portion of ***ITS*** weight, it will leave behind the energy that this light was carrying as ***ITS*** heated weight. The energy that exists as mass has to be released by the presence of ***ITS*** other way of existing: ***ITS*** nothingness.

When we look around IT

What would we find if we looked around ***IT*** using the information we have on ***ITS*** cold, clear nothingness having divine consciousness?

Imagine that you are the one who exists as a huge, clear, cold nothingness and you could do whatever you pleased with your heated, less than 1% weight while also being conscious of yourself.

Let us first accept that each of our minds is a gift from ***IT***, and whatever we discover related to how ***IT*** exists as a consciousness will not change the way ***IT*** exists. ***IT*** will even determine how long the information we find can exist. If ***IT*** wants to end our findings and do away with everything we have accumulated as knowledge, ***IT*** can simply reshape and kick us out of existence, along with our planet, our galaxy, or even the information we have sent into ***ITS*** nothingness as outer space. ***IT*** is the Commander in Chief. Everything is ***IT***.

⌘~~~~~~~~~~~~~~~~~~⌘~~~~~~~~~~~~~~~~~~⌘

****** The transferring of energy is really the reshaping of ITS heated weight within ITS nothingness******

~~~~~~~~~~~~~~~~~~~~~~~~~~~~~~~~~~~~

Let us see what we, as ***ITS*** human energy, will be permitted to discover next as ***ITS*** weight reshapes.

When I look at ***ITS*** weight as objects, I notice that I can see around an object as ***ITS*** weight. But when I try looking around ***ITS*** nothingness, I cannot see anything because ***ITS*** nothingness has expansion but no apparent beginning or end. It will be up to the scientists to see if they can detect how or where ***ITS*** nothingness, as a force, begins, or exists.

Yet while having no beginning or end, ***IT*** is not infinite, or at least ***ITS*** weight isn't. ***ITS*** weight is quantifiable, but regarding what this quantity is, we will have to wait until a mathematician figures out and tells us how there could be infinitely more nothing than something. When I receive an answer, I will post it in the databank of my webpage, **www.ricricardo.com,** for all to read. In the meantime, most believe ***IT,*** as God, to be infinite. If this is so, then infinity is based on ***ITS*** cold, clear nothingness. ***ITS*** weight would only be infinite by the infinite ways ***IT*** can reshape ***ITS*** weight.

We can see what ***ITS*** weight is doing within ***ITSELF*** because ***ITS*** weight is within ***ITS*** clear nothingness. Even as weight, as matter, ***IT*** can see what ***ITS*** weight, as fragments, is doing, since the inside of the atom is just a continuation of ***ITS*** nothingness.

Looking at ***IT*** from another angle, I can see ***IT*** as I am. First, I can look at myself because I exist as a conscious being. Second, when I look at my body, which is the same way we see ***ITS*** weight, I see my skin as an outer shell. But here is what is strange: Everything that exists, alive or inanimate, has an outer shell that is a very tiny fragment of ***ITS*** weight, and these fragments exist because of the electrons in the outermost layer of every atom. ***IT*** holds ***ITS*** weight inside with a very tiny amount of ***ITS*** weight, as the weight that the electrons carry as particles.
~~~~~~~~~~~~~~~~~~~~~~~~~~~~~~~~~~~~

Try and visualize that we, as a human body, exist as our outer shell, as a form of a speeded nothingness, because our skin is made of atoms, which have as their outer shell a speeded nothingness. And as always, ***ITS*** heated weight is inside this nothingness. Here, it would help if those who have information on how ***IT*** exists as our inner weight, as our organs and as the speed at which our brains operate, would use it to see how ***IT*** operates. For to understand, all it takes is seeing ourselves as part of this place that exists as omnipresent and that our interior is composed of ***ITS*** weight and nothingness.

Also, try thinking of ***ITS*** weight as reshaping within ***ITS*** cold, clear, conscious nothingness, that exists as an outside that does not have an ending because of ***IT*** existing as a nothingness. We can only see what is inside of ***IT***, as ***ITS*** heated weight (matter), for we cannot see what ***ITS*** other form of existing is, as ***ITS*** outer, clear, cold nothingness.

We already know that God does things in a strange way. So when you look at ***IT*** as a whole, as if you were looking to see ***ITS*** outside that extends outward, you will notice that since ***ITS*** body is a form of a clear, cold nothingness (as this Universe), ***IT*** appears to look dark, just like the darkness that now exists within ***ITSELF***, as what we see as this Universe. ***IT*** uses ***ITS*** heated weight in the form of stars to light up areas within ***ITSELF***.

Imagine again being in a dark room, and you turn on a very small flashlight, which is what ***IT*** does when ***IT*** uses a star to light up the surrounding area. What you notice is that you can see the surroundings near the light, and that when you turn on the main electric light bulb you can see everything inside the room that was totally dark. Darkness is just the absence of light, so what exists as a dark room, just like the Universe, is only dark because there is not enough light, in terms of stars, (***ITS*** heated weight) to make ***IT*** possible to see that the Universe is very clear.

As omnipresent, ***IT*** really exists as a cold, clear form of consciousness. This may seem strange, but remember this: We are made from ***ITS*** heated weight, as matter (atoms) that exists within ***ITS*** nothingness. When we think, we feel that we are conscious,

and consciousness and thoughts are not made of matter. We can think of something that is so far away that it would not be possible for that thought to get there if it had weight. If thoughts had weight it would make our thinking process slower rather than what it is now, at the speed of lightning.

I take a moment in my existence to say to ***IT***, who I once considered something that was made of something and now know is weight within 99.99% nothing, that I am grateful you permitted me, as my mind, to question your existence and see you as weight.

As weight, as matter, we can see ***IT*** and around ***IT***, and as nothingness we can see through ***IT*** and inside of ***IT***.

Looking inside ***ITS*** nothingness is the opposite of looking at and around ***ITS*** weight.

Looking around ***ITS*** weight is the opposite of looking inside ***ITS*** nothingness.

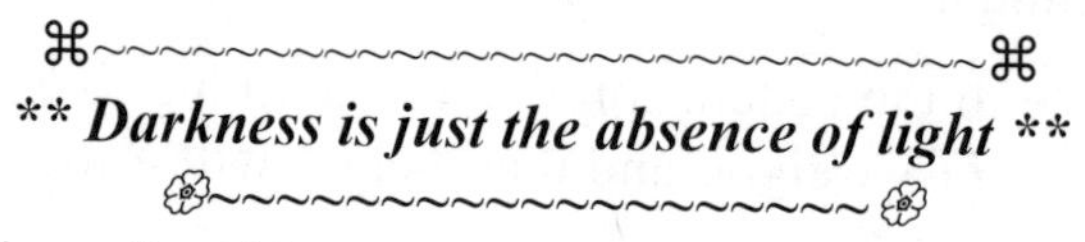
** Darkness is just the absence of light **

Our definition of nothingness

Here is something else to think about: I have asked friends what they understand when they hear or read the word “nothing”. Their answer is a natural one: Nothing is nothing, as in “not existing”.

So I have tried explaining to them that there is this nothingness, which we understand as nothing, as not existing. But there is this other nothingness that exists as this freezing omnipresent universe and as the nothingness that exists inside every atom. Believe me, this nothingness does indeed exist.

To explain this I tried to find an example, and I said that if I take an empty glass, and ask you if there is something inside, your answer would be no, as in nothing being *visible* inside the empty glass.

But I say this is not so, because inside this empty glass there are billions of fragments of ***ITS*** clear weight, that exist as the protons, neutrons and electrons of the oxygen atoms inside this glass, along

with all the other atoms of the other gases that make up the air in our atmosphere.

To think of ***IT***, those very oxygen atoms are also made in ***ITS*** own image, because these atoms also have a "shell", which is their outside, then again followed by ***ITS*** nothingness and then ***ITS*** weight.

So as you can see, I have tried to find something where nothing can exist, but found this to be impossible, because any and every thing that can exist has to exist within ***ITS*** nothingness. There is just nowhere else, no other place.

So that, even for me to think, I have to think within the existence of ***ITS*** nothingness, in the form of the matter which makes up this planet, which is also within ***ITS*** nothingness, (here I am speaking of ***ITS*** freezing nothingness as this universe) down to every heated atom that makes up my body, which also exists within ***ITS*** freezing nothingness.

This leads me to the conclusion that for anything to exist, ***IT*** must be made up of ***ITS*** weight, and that weight will always have to be within ***ITS*** nothingness.

⌘~~~~~~⌘

*** ***We try so much to find the connection to everything that exists, the connection to the Universe, but the only way we can do this is through ITS nothingness.*** ****

❁~~~~~~~~~~~~~~~~~~~~~~~❁❁~~~~~~~~~~~~~~~~~~~~~~❁

So we as humans will have to open a section in our lives, where we can place God's existence in this "place" that exists as a freezing nothingness (***ITS*** freezing nothingness, as this Universe) and this is the same nothingness that exists inside every atom.

A quote

The quote "Verily, there is an Unborn, Unoriginated, Uncreated, Unformed. If there were not this Unborn, Unoriginated, Uncreated, Unformed, escape from the world of the born, the originated, the created, the formed, would not be possible."

A friend sent me the above quote from a Buddhist treatise called the Heart Sutra, along with some commentary that I feel can help

us understand ***IT*** better if we look at the quote from a different angle or a different point of view, or at least from my way of understanding ***IT***: " 'Form is emptiness; emptiness is form. Form does not differ from emptiness; emptiness does not differ from form. The same is true for feeling, perception, mental states, and consciousness.' Here we see Buddha's original analysis of the psycho-somatic organism, but the idea is carried further. Quantum Physics has discovered that matter is nothing but a form of energy. Sub-atomic particles are merely concentrations of a field of energy that constantly appear and disappear, losing their identity as they blend into the underlying field. Emptiness is a term (also called the Void) used by Buddhists to describe the source of life, and is what Buddha called the 'Unborn, Unoriginated, Unformed.' ***IT*** gives birth to an infinite variety of forms in the Universe, which ***IT*** sustains and then reabsorbs everything -- our bodies, our minds, consciousness, nature -- is constantly being born and dying; everything is vibrations coming from the source. We are a temporary manifestation of the Void, or – in more traditional terms – we are the manifestation of the Absolute Principle. Our real nature is that of the Principle, but we identify ourselves with the appearance, with manifestation. That is why we suffer -- because we try to cling to phenomena that are impermanent. This is what Buddhists meditate on: We try to destroy the ignorance that makes us think that we are separate, substantial, autonomous beings living in a world of static, concrete entities. Thus the Heart Sutra reminds us that we must realize that the world of the senses and of our minds is only a bubble on the ocean: the Reality or Essence or Absolute Principle of the bubble is the ocean."[3]

Now, I would like to use that quote to help you, the reader, better understand ***IT.*** So I suggest that as you re-read the quote, along with my parenthetical comments, just remember that ***IT*** exists as a duality: 1) as ***ITS*** nothingness, and 2) as ***ITS*** weight.

---

[3] Available at http://www.budsas.org/ebud/ebdha141.htm. This quote is from an article called ***Why I am a Buddhist*** by Anthony Billings, Alameda, California, April 1998.

Here is how this could be said with respect to as ***ITS*** duality. All the comments in parenthesis are mine. *"Form* [or matter = ***ITS*** weight, which exists within ***ITS*** nothingness] *is emptiness; emptiness is form. Form does not differ from emptiness; [as* ***IT*** *exists as* ***ITS*** *duality] emptiness does not differ from form. The same is true for feeling, perception, mental states,* [for all these are possible only because of the way we exist as ***ITS*** weight] *and consciousness."* [These are possible because of the way ***IT*** exists as ***ITS*** nothingness.] Continuing with the next part of the quote: "Here we see Buddha's original analysis of the psycho-somatic organism, but the idea is carried further. Quantum Physics has discovered that matter is nothing but a form of energy [this is the energy that ***ITS*** weight exists as].

Sub-atomic particles [if we speak of particles, then these too have to exist as ***ITS*** weight], are merely concentrations of a field of energy that constantly appear and disappear [within ***ITS*** one total nothingness], losing their identity as they blend into the underlying field."

"Emptiness is a term (also called the Void) used by Buddhists to describe the source of life, and is what Buddha called the "Unborn, Unorganized, Unformed." ***IT*** gives birth to an infinite variety of forms [as ***ITS*** heated weight] in the Universe [within ***ITS*** freezing nothingness], which ***IT*** sustains and then reabsorbs [as ***ITS*** duality].

Everything that exists as ***ITS* heated weight** only like our bodies, and our minds. And consciousness, here consciousness is where ***IT*** exists as ***ITS*** nothingness, and nature [which is weight, because for nature to exist as something] means that nature is made of something and anything that exists as something is composed of ***ITS*** heated weight] is constantly being born and dying; everything is vibrations coming from the source. We are a temporary manifestation [as ***ITS*** heated weight, which exists within ***ITS*** freezing nothingness] of the Void or – in more traditional terms – we are the manifestation of the Absolute Principle. Our real nature is that of the Principle, but we identify ourselves with the appearance, [***ITS*** heated weight] with manifestation. That is why

we suffer -- because we try to cling to phenomena that are impermanent [***ITS*** heated weight]. This is what Buddhists meditate on: We try to destroy the ignorance that makes us think that we are separate, substantial, autonomous beings living in a world of static, concrete entities. Thus the Heart Sutra reminds us that we must realize that the world of the senses and of our minds is only a bubble on the ocean: the Reality or Essence or Absolute Principle of the bubble is the ocean."[4]

Smashing ITS weight

I would now like to discuss ideas that came to me while reading an article in the January 2007 issue of Popular Science magazine that has to do with what is known as the Large Harden Collider (hereinafter LHC). The article seeks to answer the following questions: ***1***) Why does matter have mass?, ***2***) Why does every particle have an unseen partner, and ***3***) Might the LHC produce Dark Matter and aid in the search for an extra dimension. Lastly, I would also like to comment on Jonathan Feng's statement "you never know what nature has up her sleeve".

Concerning the question "Why does matter have mass?", this is really easy to answer and understand if we remember that whoever ***IT*** is as ***ITSELF*** , or as pure energy does have weight, which is what we have been primarily focusing on, and this is natural because it is ***ITS*** weight (mass) that allows things to exist visually. ***ITS*** heated weight exists as mass and this mass can only exist within ***ITS*** freezing cold nothingness. In other words, ***IT*** exists as a duality, with ***ITS*** heated weight (mass) inside ***ITS*** constant, freezing cold nothingness, which makes up ***ITS*** shell body. In view of this, what colliders are doing is taking ***ITS*** weight in the form of protons, neutrons, and electrons and smashing them to see what may come from ***ITS*** weight as one more possibility that ***ITS*** weight can reshape into (transmutation).

Regarding "why every particle has an unseen partner", I'd like to say that we know that everything that exists as something comes

---

[4] Ibid.

from ***ITS*** heated weight, but this heated weight can only exist within ***ITS*** freezing cold, clear, transparent shell body which ***IT*** also exists as a partner to ***ITS*** heated weight. We could call ***ITS*** freezing cold, transparent shell body the "unseen partner" of ***ITS*** heated weight, but this would not be very accurate, since it is truly just the way this pure energy now exists as this freezing cold, clear universe, within which ***ITS*** heated weight can reshape (change, transmute). The mass that now exists within this freezing cold Universe cannot exist outside of ***ITS*** body, for where would this other place exist? As ***IT*** is omnipresent, there is no other place where this mass could possibly exist.

Let me clarify that I have nothing against the LHC or any other equipment that we humans can put together as ***ITS*** weight in order to understand ***IT*** better, for I too still learn from what I read, but mostly when I look at information as ***ITS*** dual way of existing. However, concerning whether the LHC might produce Dark Matter and the search for an extra dimension, I would like to repeat something that I mentioned elsewhere, that I hope will make this issue clearer: DARKNESS is really just the absence of light, for if we take anything that is dark and just shed some light on it, it will stop being dark. The only reason why our empty Universe is dark is because it would need an unimaginable number of stars to light up what now exists as darkness.

Another way to understand what we refer to as darkness is as follows: The next time you see the full moon, remember that you are seeing the side of the Moon that is receiving solar light. If you were on the opposite side of the Moon you would not be able to see it. Something similar happens when the Sun is shining and the Moon is there, but we are unable to see it. Darkness, therefore, is just absence of light. The reason why I keep returning to this subject is that as soon as you, the reader, can understand this, you will also understand that the entire Universe is ***ITS*** clear, transparent shell body, and it is not really dark. Otherwise, we could not see through this supposedly existent darkness. What has happened is that we have become accustomed to seeing things in only one way. For this reason, I will also keep asking you to return

to the photo at the beginning of this book, so that you can remember that, yes, it is one photo, the same way ***IT*** is just one, but does exist as two extremes that are one: ***ITS*** freezing cold, clear nothingness, as ***ITS*** shell body, where ***IT*** keeps ***ITS*** concentrated heated weight inside of ***ITSELF.*** This duality is what makes up our freezing cold Universe.

Perhaps you will understand this better if you remember that ***IT*** exists as omnipresent, as the empty Universe, as a cold energy in which ***IT*** moves ***ITS*** heated weight within ***ITSELF.*** Now, if what scientists call Dark Matter does exist in the form of matter was we know it, it has to exist as ***ITS*** less than 1% heated weight, and this Dark Matter would also have to exist within ***ITS*** clear, (not dark) freezing cold shell body, as pure energy, as what we understand as that now exists as this freezing cold Universe.

Now let us discuss the concept of there being an extra dimension. I say there is only ONE DIMENSION. Let me explain why: For a dimension to exist, it has to exist somewhere, so if I start with the first dimension that exists, which is the dimension that ***IT*** as pure energy has, which is the dimension that exists as the dimension that this Universe has, where everything else that exists, exists inside of this one dimension.

This dimension is composed of energy at a freezing cold temperature and we are unable to see it because it is transparent. ***IT***-God exists as a dimension that has distance, for scientists have made measurements using the speed of light as a scale. Therefore, we speak of the nearest major galaxy, Andromeda, as being 2 million light years away. Part of the problem concerning dimensions is that our minds are not made to understand things we cannot see or touch. The average human brain weighs approximately 3 lbs., but our minds are made of nothingness and every atom that makes up our body contains this nothingness which is none other than ***ITS*** nothingness. Our minds have to focus on what exists as mass, (***ITS*** weight) but at least we do know scientifically that this freezing cold, clear, invisible dimension does exist, even though our minds will continuously focus on this pure

energy's heated weight. Our minds have not found it easy to stay focused on the way this nothingness. The only way that I have found that I could be with this nothingness as ***ITS*** dimension, is through meditation. Maybe I should not really be using the word meditation in this book; what I should be using is the word connecting, for what I am really doing is just going to where ***IT*** exists and connecting to ***IT***. I cannot explained it to my mind, even though my mind knows that this nothingness permeates this Universe as a dimension, and my mind also knows that it can go to this dimension as the nothingness that I can connect to during meditation, but I cannot find words to describe this nothingness. The reason for this is that our human mind does not know how to deal with something that, even though it exists, is physically intangible. The situation is made even harder because we don't even have the vocabulary to deal with this nothingness. What I see is that our vocabulary is based on time and money, so that if I do try to describe this nothingness as something, I could then be describing ***ITS*** weight. I could perhaps use the word "faith" to describe ***ITS*** freezing cold, clear, invisible way of existing, but I will refrain, because there is no need to metaphysically explain something that is known scientifically to exist. The Universe is omnipresent.

We will have to hold off until more scientists are funded to find ways of understanding this pure energy that is there as the dimension that now exists as this clear, transparent, freezing cold nothingness that exists in between the fragmented portions of ***ITS*** heated weight, or until we find how to use this freezing cold energy for economic profit. Maybe if we could find out how to bring this freezing coldness back to Earth to replace the energy we have to spend on refrigerants, someone with monetary energy will focus on this freezing cold energy as something. Anyhow, it is up to ***IT***, whether we are allowed to know and understand how ***IT*** exists as ***ITS*** freezing cold, shell body for ***IT*** can reshape (destroy) all of us by just taking ***ITS*** weight, and throwing it on us as a huge asteroid, that could knock our whole planet out of existence. But as I have mentioned elsewhere, I feel that ***IT*** will take some of us somewhere else before our planet's death date arrives, which, by

the way, is estimated to be more than billions of Earth rotations away. Who knows? We just might be permitted to escape the last black holes consuming each other, just before all black holes become just one total black hole, which will bring us back to the moment of a new big bang as a new possibility as ***ITS*** heated weight.

I have considered writing a book just on this, but I would need some one that could participate in this venture, in putting together a book for I have already out lined the information that would be used. It is the filling in between that needs attention for a book of this nature to take place, for someone with a creative science fiction mind. But then again, that is up to ***IT***. For all I know ***IT*** might send someone to help me, or ***IT*** might do it as someone else, and then again, ***IT*** may never let any thing happen concerning putting this subject on paper. On the other hand, since ***IT*** is always searching for all existing possibilities, ***IT*** might let it happen.

Going back to our discussion about dimensions, let me give you, the reader, an example. When we use the word dimension, we are using it to describe the way something exists, usually something that exists as matter, which is really ***ITS*** weight. But we forget that this dimension that this something has, is only possible because of ***ITS*** nothingness, specifically the 95%-99.99% nothingness that this something exists as in terms of atoms, which is what is giving this something a dimension. As you can see, what our minds can focus on is ***ITS*** weight, not ***ITS*** nothingness, even though it is ***ITS*** nothingness what makes this something possible.

⌘~~~~~~~~~~~~~~⌘~~~~~~~~~~~~~~⌘

******Help wanted: Someone to invent an invisible ruler, that can measure something that exists, but exists as an invisible dimension known as the empty Universe******

✿~~~~~~~~~~~~~~~~~✿~~~~~~~~~~~~~~~~~✿

Let's continue in this vein: When we see a one meter steel cube, what we are really seeing is a surface made up of ***ITS*** speeded nothingness, that has a minute amount of ***ITS*** weight forming the electrons that surround the atoms that make up that surface. This surface is followed again by ***ITS*** nothingness, as the 95%

emptiness that each one of these atoms contains inside itself. This is what makes up the cubes outer surface which has measurable dimensions: height, width, and length.

Let's suppose, for the sake of this discussion, that this steel cube weighs 1000 lbs, (as ***ITS*** heated weight), and is occupying 100 square feet. We know scientifically that this 1000 lb. steel cube is made of atoms, and all atoms exist having electrons as their outer or outside layer. The electrons are what ***IT*** uses to hold ***ITS*** fragmented weight inside the atom (protons and neutrons), which is then followed by 95% empty space. The innermost part of the atom is where the protons and neutrons exist, containing the bulk of the atom's weight. Now, if we take these 1000 lbs as the 100 sq. ft. dimension as the way that this steel cube exists and we remove the 95% nothingness that exists inside each one of the atoms that make up the cube, we would still have 1000 lbs but this 1000 lbs would not occupy 100 sq feet as a dimension. It would occupy a smaller space; maybe 5 sq feet, as a dimension. Now, our minds won't be able to see this 5ft dimension in this manner because we can only see an object's dimensions when it is existing as ***ITS*** weight in the form of matter, because our minds still have a hard time focusing on ***ITS*** nothingness as ***ITS*** one total dimension. This is the only dimension there is: This empty, freezing cold Universe which is ***ITS*** shell body which houses ***ITS*** heated weight that ***IT*** reshapes in to everything that we perceive as matter.

Another reason why it seems hard to understand how ***IT*** exists in ***ITS*** totality is that our language is not made to talk about things as just existing, without having to use the word time. Let me show you what I mean: How can I write you about the way ***IT*** exists as omnipresent, that is, being in all places at the same moment, where all ***IT*** is doing is reshaping ***ITS*** heated weight, but not as time? If we try to see what ***IT*** is doing to ***ITSELF*** as ***ITS*** heated weight, as omnipresent, as just one moment (here I again I have to use the word moment to try and describe to you how ***IT*** exists!) which is not a moment in terms of time because this is just the way ***IT*** exists. One could say ***IT*** just is, for in ***ITS*** totality, ***IT*** is timeless. So trying to visualize ***ITS*** reshaping as one continual moment,

without having to use our mechanical time system has not been easy, but it is possible, and when you have adjusted to seeing ***IT*** as just existing without using our mechanical time system, you will then begin to understand ***IT*** better, and you will also understand yourself better when you realize that we too have always been existing as just one moment of ***ITS*** timeless existence.

If you find it hard to conceive of something that is timeless or that exists in a timeless state, remember that our human bodies are finite. Our bodies are born, grow, live for a time, and then die. In other words our bodies have a beginning and an end. Our ancestors understood this when they were still living in caves. They also observed that there are rhythms in the natural world, and started keeping track of them: days and nights, the four seasons, the lunar cycle, the positions of stars in the night sky. This was the beginning of our mechanical time system which we use, among other things, to make measurements between the beginning of our body's life and its end. ***IT*** on the other hand, has no beginning or end, so ***IT*** has no need for time. ***IT*** is timeless. Therefore, for ***IT***, it is always NOW.

In the same way that our bodies are finite in time, having a beginning and an end; our bodies are finite in space. Our bodies are limited and separated from other things by their skin. ***IT*** is infinite. ***IT*** has no beginning or end in spatial terms. ***IT*** is not limited, has no boundaries, and is omnipresent. There is no other place but ***IT***, and the Universe we know is commensurate with ***IT***; one and the same. Therefore, for ***IT***, it is always HERE.

The only thing that changes in the Universe is ***IT*** reshaping ***ITS*** weight, as matter, within ***ITSELF.*** We have been gathering information on how ***IT*** exists from as far back as the Big Bang, up until where we are now, as ***IT*** has reshaped ***ITS*** heated weight within ***ITS*** cold shell body as ***ITSELF.*** But not as time, for ***IT*** does not exist as time. ***IT*** just exists! So try to look at what is now taking place with in this Universe as the way ***IT*** is reshaping ***ITS*** heated weight within ***ITSELF*** as omnipresent, but not in terms of our mechanical time system, for ***IT*** has always existed. ***IT*** is now.

IT was long before the Big Bang. And ***IT*** will continue to be forever and ever, for ***IT*** is totally independent of our human invention of our mechanical time system.

Let me mention that it is not easy to see this nothingness as one total dimension, but yes, it is possible to visualize this nothingness when you look at the Universe and remember that what you are seeing is ***ITS*** weight inside of ***ITSELF*** as ***ITS*** freezing cold, clear, transparent nothingness that has measurable physical dimension and exists as an Omnipresent pure energy, for it is not nothing as nothing, and most of all, it does exist as a freezing cold temperature which has the opposite quality of ***ITS*** concentrated heated weight.

But let me go back to the way this one dimension exists, where everything that does in fact exist must be contained. This is a simple concept: What we understand as this empty Universe, is just this one dimension existing as just one huge, freezing cold, empty, clear nothingness that has expansion or extension and can be measured in terms of distance. However, our minds are focusing on the distance that exists from one fragmented portion of this pure energy's heated weight to another portion of ***ITS*** heated weight.

⌘~~~~~~~~~~~~~~⌘~~~~~~~~~~~~~~⌘

There is only one dimension, and that is the dimension that ITS body exists as, but IT is an invisible form of a freezing cold energy

❀~~~~~~❀

And let me also mention that this dimension does not exist as a time dimension. Let me set up another example: Imagine that we can go back to when ***IT*** had all ***ITS*** heated weight in just one place within ***ITS*** nothingness, as in the moment just before the Big Bang. There would be no way that we could use our mechanical time system to measure distance, because there would be no distance in terms of ***ITS*** one total weight, for we can only use our mechanical time system to measure from one portion of ***ITS*** fragmented weight to another portion of ***ITS*** fragmented weight. We cannot apply our mechanical time system to measure how ***ITS*** huge, freezing cold nothingness exists, because we have no reference point from which to begin a measurement. This is why

we refer to ***IT*** as having no beginning or end. Another reason why this dimension that exists as ONE is timeless is because this belongs to ***IT*** as ***ITS*** nothingness body. Our mechanical time system can only be used after the Big Bang, which is when ***IT*** fragmented ***ITS*** weight, so that we could exist along with our mechanical time system to measure from one fragmented portion of ***ITS*** weight to another, using our time system.

Finally, I would like to remark on Jonathan Feng's statement that "you never know what nature has up her sleeve". As I have said elsewhere, I find it more helpful to use the word God instead of the word nature because nature is not a "she" in the sense of being of female gender and at the same time, I can understand nature better when I remember that nature is really ***IT*** as God, or pure energy. Neither God, nor energy nor nature are words to which gender can be applied because gender properly refers to the a reproductive polarity of biological bodies. Seeing that we cannot speak of God, energy or nature as "he" or "she", I would like to rephrase Mr. Jonathan Feng's remark as follows: We never know into just what ***IT*** is going to reshape ***ITS*** heated weight as ***ITS*** one moment of ***ITS*** existence.

IT as one nothingness

IT exists as one total nothingness, as ***ITS*** shell, which behaves as an inner and outer shell simultaneously, and ***ITS*** weight is in the inside of the shell. Life is a part where ***IT*** has taken ***ITS*** fragmented weight that exists within ***ITS*** total nothingness and has given this fragmented weight mobility and an appearance in order to exist as something, by placing a tiny amount of ***ITS*** weight as the weight in the electrons, protons, and neutrons, so that ***ITS*** weight could be visually seen, as matter, so that when we see ourselves as something that is alive, what we are really seeing is ***ITSELF***. If I start by what I see as something, let's say a living person, what really exists is this: first, the person is there because ***IT*** took a tiny amount of ***ITS*** weight and attached ***IT*** to ***ITS*** ***MAXX-SPEED***(which you will read about later on in this section) so as to become electrons. ***IT*** then took another, larger quantifiable portion of ***ITS*** weight, and surrounded this bigger weight which

we call protons and neutrons, with what ***IT*** had previously reshaped into as electrons, so that what we call matter could exist, with a given distance as the nothingness, the 95% empty space that exists inside the atom.

Now, if we go back to seeing the person that is alive again, what is happening is that the reason why we can see a person as being physically there is due to the fact that ***IT*** took a very tiny amount of ***ITSELF*** as ***ITS*** weight to become the electrons that exist as the person's outermost layer, which is still alive, and then ***IT*** then formed ***ITS*** other fragmented weight into the protons and neutrons that make up the atoms that compose the person's body, so that when we see this person, we see him/her because of the way ***IT*** is using ***ITS*** weight(matter).

Now let me return to ***IT***S nothingness again and go back to the concept of illusion, to just what it is that is driving these electrons with ***ITS*** way of existing as a high speed (***MAXX-SPEED***) around ***ITS*** other weight as protons and neutrons. Now here is what is very interesting: ***IT*** is doing this within ***ITS*** total nothingness, so that when you see a person, what really exists is that ***IT*** took that very small amount of ***ITS*** weight as electrons, gave ***IT*** distance or separated ***IT*** apart from that other part of ***ITSELF*** that exists as ***ITS*** fragmented weight as protons and neutrons in order to form atoms, and gave ***IT*** what ***IT*** already exists as. Not life, but as a divine consciousness, which ***IT*** exists as, as the nothingness that is always a constant.

So, returning to us being alive, this is the condition in which ***IT*** gave ***ITS*** once total weight (the weight that existed at the moment of the Big Bang) and fragmented this weight, as electrons, protons, and neutrons, so that ***IT*** could reshape into the matter that we exist as. In addition, ***IT*** gave this weight certain functions, like our brain, heart, liver, lungs, and certain properties so that this weight could have mobility, so that ***IT*** could exist as ***ITSELF***, as a divine consciousness, which we call being alive.

Again, when we see humans, all of these humans and everything else exists within ***ITS*** one total nothingness, so that even if we get the impression that we are outside of something, we are really still

inside of *ITS* total nothingness, and the reason is that *ITS* nothingness exists in everything that exists, such as air, water, trees, houses, trains, planes, and every other thing on and in this planet; all is inside of *ITS* total nothingness, because anything that does exist, in order to exist, has to be made of *ITS* weight which has to exist within *ITS* one total body that exists as what we see as this Universe.

Returning then to what I started out to say, *IT* is very interesting to see things as being out there, in what we call reality. Nevertheless, everything out there is really inside of *ITS* one total nothingness.

All of this will make more sense if you always remember that nothing is really ever created or destroyed, for everything that has ever happened or will happen, has always been and will always be *IT* as one.

You might understand *IT* better this way: It is the separation of *ITS* fragmented weight that give us distance (or extension in space) within *ITS* one total nothingness, which gives us the illusion of there being trillions or googols of things existing out there, when in reality, there is just one of *IT*, where all *ITS* fragmented weight is moving within *ITSELF*.

Light before the Big Bang

Was there light before the Big Bang? I said no before, but perhaps there was. Light is the result of a positive and negative coming together. Just before the Big Bang, when *IT* had all *ITS* heated weight in one place as dense matter, *IT* existed as two forces: the huge, cold, clear nothingness as *ITS* outer structure, and *ITS* heated weight, which produces the visual effect of something existing as matter. The heated weight is the opposite of *ITS* cold, clear nothingness.

Since *IT* does exist as two opposite forces, and at the moment before the Big Bang *IT* had all *ITS* heated weight in one place, *IT* could mean that this very dense, heated weight was in constant contact with *ITS* opposite force of cold nothingness. This means that at least on the edges of where *ITS* heat existed as a positive force, there was contact with *ITS* cold negative force.

⌘~~~~~~~~~~~~~~~~⌘ ⌘~~~~~~~~~~~~~~~~⌘

***Life is where ITS weight exists with mobility as an outer shell, governed by IT, as ITS divine consciousness as a nothingness** *

❁~~~~~~~~~~~~~~❁ ❁~~~~~~~~~~~~~~~~❁

Imagine that you could visit this first stage of ***ITS*** existence. I say imagine because when ***ITS*** weight was in one place, we could not exist; for we, as matter, came after ***IT*** fragmented ***ITS*** weight into atoms. But if you could and you did see light, this would likely be because ***ITS*** heated weight, as a force, was making contact with ***ITS*** opposite force, the invisible cold nothingness, in the same way the opposite forces in a battery, produce electric current that can power a light bulb, when the opposite poles make contact. As light, ***IT*** transfers a dual energy. Here again we find a way to understand what ***IT*** can do as pure energy, as a way of existing. Important here is that you use the information to better understand the relationship between you and ***IT***.

⌘~~~~~~~~~~~~~~~~⌘ ⌘~~~~~~~~~~~~~~~~⌘

*****We are IT totally, as ITS weight, and as ITS nothingness*****

❁~~~~~~❁

Can IT see?

To see something, I have to use my eyes. And this is how I know that ***ITS*** weight exists as the matter the Universe contains. We see visually in this stage of ***ITS*** reshaping because of ***ITS*** weight, but we know that before this Universe came into existence there was a stage, which I refer to as stage one, where ***IT*** existed as the same nothingness that is now our outer space, where ***IT*** had all ***ITS*** weight in one place within this space. We refer to that as the moment just before the Big Bang.

While ***IT*** did exist as this stage, ***IT*** could not see visually because what we call eyes did not yet exist. So ***IT*** could not see. ***IT*** does not need eyes to reshape. ***ITS*** higher quality is in ***ITS*** knowing, as in how ***IT*** knew to reshape into the orderly atoms that now exist as ***ITS*** quantifiable weight. We understand this as the weight of the protons and neutrons in each atom's nucleus, which ***IT*** keeps separate from other fragmented weight with the use of speed. We know this speed exists in the form of the electrons in each atom

and we also know that ***ITS*** consciousness or intelligence is in the nothingness that ***IT*** exists as within each atom.

As to what ***ITS*** weight can or cannot do, try to follow this: This same nothingness is where ***IT*** knows what ***IT*** can do with ***ITS*** weight as atoms and what ***IT*** can reshape into within the nothingness--the nothingness that exists outside the atom and the nothingness that we see as this Universe. And ***IT*** knows what ***IT*** is doing with ***ITS*** weight inside the atom as the nothingness that ***IT*** exists as.

So you see, ***IT*** may be more of a doer than a seer when it comes to reshaping. ***IT*** does not need to see what it is doing, for it is doing it within ***ITSELF***, and there is no chance of it losing ***ITS*** weight outside of ***ITS*** total nothingness.

I know that what I offer you is all imagination, but as I have said before, any qualities that we have as humans come from ***IT*** as a way of ***ITS*** existence. When I question any qualities that I may have, I know that ***IT*** placed them there. Even our eyes are the result of the weight that ***IT*** exists as. You might see it better this way: ***IT*** is using ***ITS*** weight as our bodies to see and understand ***ITSELF*** better.

⌘~~~~~~~~~~~~~~~~~~~~~~~~~~~~~~~~~~~~⌘

***** IT sees through the eyes that we have, that IT exists as us.*****

✿~~~~~~~~~~~~~~~~✿ ✿~~~~~~~~~~~~~~~~✿

I hope I do not bore you. I simply find it more interesting to focus on ***IT*** as ***ITS*** total omnipresence, as in ***ITS*** weight and nothingness, than any of the humans that ***IT*** exists as on this planet.

I do not dare to say that ***IT*** is fat and needs to diet, but we now know that ***IT*** carries ***ITS*** weight in an area that is empty and clear and does this in order to move at high speed, to spin, and reshape. ***IT***, as less than 1%, weighs millions, billions, trillions, or more tons. And ***IT*** uses color to see that which ***IT*** reshapes into, since what ***IT*** reshapes into is empty. But what if we could remove emptiness?

We are IT

Because we are made of the atoms which are made from ***ITS*** pure energy, we are ***IT***. We are not all of ***IT***, however. We, as atoms, are actually fragments of ***IT*** as ***ITS*** weight. We are the weight that exists in protons and neutrons, and we are also ***IT*** as the speed of all the electrons that permit an atom to exist. And we are also ***IT*** as ***ITS*** consciousness: the clear nothingness that ***IT*** exists as. And so we are God, as tiny fragments of the different qualities that ***IT*** exists as: the weight, speed, nothingness, and energy. However, we are not the God that exists as the pure energy that exists as one, as a whole. As much as I might say that we are ***IT,*** as ***IT*** exists as ***ITS*** fragments of weight, speed, and nothingness, we are not the God that is running this Universe (***ITS*** total weight and nothingness combined).

Constants

Here is an idea that is related to an article that I was reading in the June 2005 Scientific American, entitled *Constants* by John D. Barrow and John K. Webb. They reminded me of the way ***IT*** exists as constants, such as electrons and the speed of light.

I need to mention that I still learn from reading things, but when I read an article like the one above, I have to question it. I am not disagreeing with John D. Barrow and John K. Webb. It just makes me think of some of the ways that ***IT*** exists as constants, such as the electron and the speed of light. This is the reality of ***ITS*** present existence in this stage.

I say this because as we continue our investigations into the constants that now exist, we should remember that these constants are also ***IT*** at this stage of ***ITS*** existence. I say this because I also have to see the things that ***IT*** exists as in its different stages so that I can see what the constants are in all of ***ITS*** stages. For example, when I view the electron, or the speed of light as being a constant, I have to remember that these two, just to mention a few, may not have existed in ***ITS*** pre-Big Bang stage. This may be because electrons or light are carriers of ***ITS*** weight, and before the Big Bang, ***IT*** had all its weight in one place. The only constant I think

was there then and continues to be there now is ***ITS*** huge, cold, clear nothingness.

When it relates to constants, I have to say "thank you" to ***IT***, for ***ITS*** constantly existing. Imagine what would happen if pure energy, or God where not constant enough to continue existing.

IT as cyberspace

I am no expert on cyberspace technology, so if you have more to offer, please send it to my e-mail address located in the back of this book, so I can post it on my website databank.

When we send information electronically, as fax or e-mail, we are sending something, although not physical, as information. Let us say that I am going to send you an encyclopedia or a huge dictionary as an e-mail. This is a lot of information, and we know that the encyclopedia of information does exist, so we have to say that it is something. As something, it has weight attached to it and can be called matter, or more specifically, we can call it electrical material. My curiosity concerns the weight of this electrical material.

I asked a computer savvy friend what he knew about this subject of electronic information transmission. According to him, information travels through the internet as a 1(one) and as a 0(zero). This is called the *binary* system.

To me, the word or number one [1] signifies the existence of something. And when a zero [0] is placed after the one, as in 10, ten, there is more than one, and when a zero is placed before the one, there is less than one; that is, the one is fragmented into smaller pieces. Additionally, zero could stand for something that does not exist, as in "how much money do you have?" "Zero." And the many other ways that we understand zero as nothing. And you can guess it's this nothingness that interests me more than the something that exists as an encyclopedia. Yet, I have to remember that 0 [zero] is a symbol or representative number that stands for nothing.

Because ***IT*** exists as a nothing, as in ***ITS*** huge 99.99% nothingness, I question this 0 (zero) that exists as e-mail. I question

its meaning —zero as nothingness —in the many aspects of the world we live in. Actually, I question everything that exists as something. If this something, let us say cyberspace, or anything else that can exist as something, has ***ITS*** weight, then where is ***IT*** and what is ***IT*** doing with ***ITS*** 99% speeded nothingness? ***IT*** has to be there as something, which in this case is represented by the number 1.

You can see why I question how ***IT*** operates as cyberspace, which is something that exists yet cannot be seen. You can also understand that the reason we cannot visually or physically see cyberspace is because of ***ITS*** high-speed nothingness that cyberspace exists as and ***IT*** has so much of. And based on our previous conclusions, we can also see that ***IT*** placed a tiny amount of weight on this speeded nothingness, which is why cyberspace also exists as the encyclopedia. ***IT*** produced electrons and then used them, with ***ITS*** tiny weight, to encircle more of its weight with quantifiable fragments, known as protons and neutrons. In this manner ***IT*** produced the atom. From the atom ***IT*** reshaped into stars so that it could send light out into the rest of the 99% nothingness that ***IT*** exists as. This is how we can see that ***IT*** is sending out ***ITS*** very tiny, fragmented weight in ***ITS*** high speed nothingness to the rest of ***ITSELF***. And ***IT*** knows what ***IT*** was doing with ***ITSELF***, for we know that this Universe exists in a very orderly and perfect way. We and the Universe are an incredible miracle that comes from ***ITS*** reshaping. What's more, cyberspace could not exist if ***IT*** had not reshaped into us.

Thus ***ITS*** weight is making cyberspace possible. This is because everything, from you to the computer, operates with ***ITS*** high speed electricity that travels through ***ITSELF*** as the equipment transmitting and receiving ***ITS*** high speeded nothingness as data, as electrons that are traveling near ***ITS MAXX-SPEED***, which is made slower by carrying the minute amount of ***ITS*** weight that exists in the form of electrons.

But like I have mentioned before, I have no knowledge of this area of ***ITSELF*** called cyberspace. I can only say that when we are using the word *space*, we are referring to ***ITS*** nothingness, not so

much as ***ITS*** weight, and in this nothingness, ***IT*** exists as the 99% space that is an invisible shell.

So again, let us see who is out there that has more information on cyberspace. This is such a totally new area that I feel we are just opening the door to it. And what we know is just the beginning of how ***IT*** will reshape as this new area, cyberspace, which we cannot visually see, but we can intellectually understand.

So, continuing with e-mail, which travels not only as zeroes (0) but also as ones (1), the ones signify the addition of weight, as things, as in one car, two cars, one person, two persons. We need to see it this way, otherwise we could not deal with the way we now exist, for we cannot for example, purchase zero of something. But our tallying is really just a way of adding fragments of ***ITS*** one total weight. Another way you might understand this is that we are adding up fragments that are to the *minus* side of the number line representing ***ITS*** total weight. If we could gather all ***ITS*** weight that now exists within ***ITS*** nothingness, as the existing Universe, that total would then be the less than the 1% of the Universe ***IT*** has as weight.

So when we send an encyclopedia by electronic mail, we first have ***ITS*** weight as: (1) our person as weight, (2) the computer as weight, and (3), just as important, the electrical energy that is required to send the encyclopedia as electronic material as ***ITS*** weight. This encyclopedia exists as zeroes and ones, and whoever receives this encyclopedia will be seeing something, but not as the weight that the encyclopedia printed, for I am aware that the weight of a physical encyclopedia exists due to the weight that the paper and ink have as atoms. The encyclopedia, which is being viewed as an e-mail encyclopedia, exists because of the weight that existed as electricity that was needed to send the encyclopedia as electronic information from one point to another. We also needed ***ITS*** weight, as the computer, and the electricity to view it. When I view my e-mail, I am aware that this is only possible because I am using ***ITS*** weight, as my computer, and more importantly, ***ITS*** high speed nothingness transferring ***ITS*** weight, as the weight that

exists in the electrons that create the electric current that we call electricity.

Here is something else to remember: As we send information on the internet, the information, as energy, is made of ***ITSELF*** as pure energy. This means that ***IT*** knows all the information we send. ***IT*** is all-knowing. Let me also mention that this world of cyberspace is something that came from ***ITS*** reshaping as humans.

When ***IT*** existed without cyberspace, which it did before the Big Bang, ***IT*** knew everything that existed within ***ITSELF*** as its weight. And before the Big Bang, it was easier for ***IT*** to keep track of ***ITS*** weight because all ***ITS*** weight was in one place as a very dense singular point. After the Big Bang ***IT*** has had to keep track of ***ITS*** weight as the celestial bodies that now exist and now too, as the zeros and ones that exist in this cyberworld, which is at its beginning.

And, ***IT*** has already taken this system of zeros and ones outside of our planet and into ***ITS*** huge nothingness, as outer space, as the rockets and satellites that now exist out there. ***IT*** has also taken this system farther out, as far out as the *Voyager* will travel within ***ITS*** cold, clear, speeded nothingness as zeros and ones.

Prior to zeros and ones and computers, ***IT*** was all-knowing, as the information that existed in books. But now, as zeros and ones, ***IT*** has more and faster access to ***ITSELF*** because electronic information uses very little of ***ITS*** own weight. The electronic encyclopedia, for example is not very large, for the electrons are made of a very small and invisible form of speeded nothingness that exists in a world within a world when compared to our world, which is just the size of a grain of sand compared to ***ITS*** hugeness as ***ITS*** nothingness. Let us see if there is a reader that knows the weight of these zeros and ones, for these zeros and ones do exist; therefore they have weight attached, perhaps similar to the weight of any subatomic particle.

Let us recall that ***IT*** has always existed as this nothingness where ***IT*** exists as a divine consciousness. Now, ***IT*** is using this nothingness as a cyber world where ***IT*** is reshaping itself within

this nothingness of zeros and ones as information on ***ITSELF***. ***IT*** will have all this information on ***ITSELF*** even after we and the many that will come after us are long gone. Since ***IT*** still has information on ***ITSELF***, as all the information that exists in libraries, or as all the information that exists on paper, ***IT*** now has more of it in the form of zeros and ones, and maybe someone will see more of what ***IT*** is doing as these zeros and ones. This someone might one day decide to print it, but by then we may not be using paper or ink, as ***IT*** continues reshaping ***ITSELF*** in this miniature area of zeros and ones. We have some information on ***ITS*** miniature weight, which exists in the subatomic world as the particles that make up protons and neutrons. But this subatomic world does not have that which ***IT*** has as this cyber world, which is that ***IT*** can communicate with the rest of itself as information, unlike subatomic weight which stays confined within the existing unit as an atom.

As for sending this as ***ITS*** weight with these zeros and ones, we can send text as alphabet, photos and graphics, but for now, as far as I can see, we cannot send things like food or cars.

And these zeros and ones are great to use as computer text, and to store and move around without the need of ***ITS*** weight as paper and ink. This is the way the alphabet looks in the binary system, as samples of the letters, a, b, c, x, y, and z. But why do they consistently begin with a zero? I think zero comes first because our number system starts with zero and then one. I understand that to be ***ITS*** way of understanding ***ITS*** weight that exists in the atom, and as the weight that may exist as these zeros and ones.

a	01100001	X	01011000
b	01100010	Y	01011001
c	01100011	Z	01011010

Maybe there are experts out there that can add up a million letters in binary and let us know their total weight. Or maybe they can calculate whatever amount of letters is necessary to give a specific weight such as one gram. This would give us something to help us understand how much of ***IT***, as minute weight, is being used to

operate this cyber world. We could compare it with the real physical world. What, for example, would a cyber encyclopedia weigh?

I have tried to understand ***ITS*** weight as zeros and ones, but have reaped few results. I'm resigned to accepting that the binary system of zeros and ones simply works as cyberspace, and that I may never understand ***ITS*** weight in this mechanical area, except to say, anything that could reshape into this Universe would not have a problem understanding ***ITSELF*** as this tiny world of zeros and ones. But here again, it makes no difference to me how much of ***IT*** exists as cyberspace, for whatever this number may be, it will not change the way ***IT*** is. It will only help us understand what ***IT*** is composed of, and how ***IT*** operates as ***IT*** reshapes ***ITS*** heated weight. ***IT*** may not make sense to us, but it does to ***ITSELF***, for all is happening within ***ITSELF*** as ***ITSELF***, as the new things that can take place from the reshaping of ***ITS*** weight. ***IT'S*** incredible, isn't ***IT***?

Head to toe

As far as ***IT*** being in all places at the same moment, let me just add a few things. The closest I can figure ***IT*** being everything is in ***ITS*** nothingness, which is everywhere. Because of this nothingness, I can see ***IT***.

From the top of our heads to the tip of our toes, and as far as the Universe extends, this nothingness is there in the atoms with their speed and rotation. This nothingness composes more than 99.99 percent of the Universe. It is there that ***IT*** moves ***ITS*** weight, where ***IT*** exists as a total. The difference between my hair and my toes is in the density, and arrangement of the atoms; not in the emptiness within them.

If I think about ***IT*** as being everywhere, I can see that the one thing that is everywhere is this nothingness; from every planet that may exist down to you and me. Even that thought is from the nothingness that ***IT*** exists as.

How does ***IT*** permit me to think, and you and everyone else at the same moment, and be at the other end of ***ITSELF*** as this Universe

at the same time, and know what is happening? I apologize. I have tried to understand, but I have no answer from ***IT*** yet as to how ***IT*** is an omnipresent thought.

Maybe the answer has to come from the many readers that may have seen what I have not yet been able to see. So think about this: We can think, which means that ***IT*** can also think, we being in each other's image and all.

I have thought: How did ***IT*** think as to what was needed for ***IT*** to become, and in such perfection, as all the atoms that populate the nothingness of the Universe? So if you think that we can think, imagine how ***IT*** could think about how ***IT*** was going to reshape ***ITS*** heated weight and become every perfect atom that now exists, within ***ITS*** own nothingness.

Let us look a bit further into this nothingness. Protons can become infinitely weightless; which is what happens when a proton decays into smaller fractions that become mere waves of energy.

Think: How did ***IT*** know how to divide ***ITSELF*** as a given quantity of ***ITS*** heated weight to make the proton in each and every hydrogen atom? This takes very precise thinking, which may be easy for ***IT***, but we are only babies as thinkers and planners. We may need higher and higher speed computers to help us understand better how ***IT*** is and how ***IT*** operates.

Peace and violence

Let me offer you one of my off-the-wall thoughts. Matter is where ***IT*** has heat, and heat is very violent; this is why a volcano turns heat into something new; this is the nature of ***ITS*** heated weight. We would not be here if it were not for ***ITS*** ever changing weight, which we can see as far back as the moment of the Big Bang as a very violent explosion. ***ITS*** weight will keep colliding with ***ITS*** other parts as weight in order to continue ***ITS*** search for all existing possibilities to reshape into.

The opposite of ***ITS*** heat is coldness, which could also be where this peacefulness could exist. When someone, as the saying goes, is heated up, he or she is near a violent stage of exploding, as in being "out of control". And when someone is keeping his or her

cool, this person is more in control and at peace. It is in ***ITS*** nothingness where ***IT*** exists in peace. If there is nothing as something, which is needed for violence to exist, then what we understand as peace, (which is just a feeling, for peace does not exist as something, like an object), must have its roots in the nothingness.

Some people know they can find this peaceful feeling in meditation, which is connecting with ***ITS*** nothingness. Peace is also found in prayer, which is also a form of connecting with ***ITS*** nothingness. We should look for as many ways to connect with ***ITS*** peaceful nothingness as we can. We have tried to find peace in the things that ***IT*** exists as, things with weight, but this strategy has not been very effective.

A gift for us

Here is something that I consider a gift from ***IT.*** But before proceeding, I must say that I find that no other creature other than humans have been given this gift, for as you will see, animals, like cats and dogs, just to mention some, cannot understand this gift that we have.

And the gift that we have is this: That we have been given the opportunity to look inside of ***IT*** as ***IT*** exists.

Let me make this just a little clearer: We have, from the moment of our arrival on this planet, been focusing on ***IT*** as something that exists as being physical, or as something we can see, and we have understood ***IT*** more as existing here on Earth, but we are now more informed, as to ***ITS*** way of existing, especially as pure energy; as ***ITS*** nothingness and as ***ITS*** weight; as ***ITS*** housing (the existing empty Universe). Now to make this easier for you to understand, there are a few things that we have to remember: Since God made every thing that exists, (as ***ITS*** weight), then all the matter that exists in what we call outer space, is ***IT***, as ***ITS*** weight, that ***IT*** has within ***ITSELF***. This being so, then think about this: As we look inside this existing Universe, we are really looking into ***ITS*** invisible body, that exists as ***ITS*** housing in the form of a cold, clear frame that exists as a nothingness, where ***IT*** can move ***ITS*** weight around within ***ITSELF.***

So you can see why ***IT*** is a gift from ***IT*** to us, being permitted first to exist as humans that exist as ***ITS*** weight, and also with the eyes ***IT*** reshaped into, so that we could use the mind that ***IT*** produced from ***ITS*** reshaping, so that we could be able to see things as ***ITS*** weight, that exists with in ***ITS*** nothingness as this Universe. This is a gift that only humans, among all the other creatures on this planet, have.

And I know that for me, ***IT*** was a little hard at the beginning to accept that I was looking into ***ITS*** inside. This took some adjusting, but as soon as I accepted that if everything that exists is ***IT*** as omnipresent, be ***IT*** under the name of God or pure energy, ***IT*** then became easier to see that when we look at what exists made from matter, be ***IT*** as atoms, or as what exists in what we have been referring to as outer space, it is clear that there is no outer space, for this outer space is really ***IT*** as omnipresent, ***IT*** as ***ITS*** interior. So now, when I look into what I once thought was outer space, I am aware that I have to remember that I am really looking inside of ***IT,*** and that what I see inside of ***IT*** (this Universe) as celestial bodies is really ***ITS*** weight, which we refer to as atoms or matter.

So I have to remember that first, ***IT*** permitted me to exist as ***ITS*** weight, surrounded and governed by ***ITS*** nothingness, so that I could use what exists outside of me as information, such as there being a Universe, so that I now view what is taking place inside this Universe as really ***ITS*** weight moving about within ***ITSELF***, and this to me is a gift that is precious; not priceless, for this gift has nothing to do with money, ***IT*** has to do with only me, and ***IT***, in allowing me to understand ***IT*** just a little more.

So I say: "Thank you, (***IT***), for permitting me to see how you move your heated weight around your freezing cold, transparent nothingness and for letting me know that when I now see you transferring energy, what I am seeing is the way you transfer your heated weight, within your constant, as the never changing consciousness that you exist as, which is a form of nothingness, so that now I know that you (***IT***) are only constantly transferring your

energy as your heated weight within your never-ending, constant, cold, clear nothingness."

Let me mention, that to some people the above information may seem odd, but we know that many things that were once said seemed unacceptable to others, such as our planet being round, or that our planet travels around the Sun, or our planet being the only planet to exist in this Universe. So if I say that we may be the only ones that have been given this gift of knowing that ***IT*** is from our planet that we can see into ***IT*** as ***ITS*** invisible frame, or ***ITS*** body that is composed of a cold, clear nothingness, is, at least to me, a gift.

ITS Strangeness

I would like to add some more information on ***ITS*** strangeness, and ***IT*** would help if we remember that ***IT*** is strange in ***ITS*** way of existing, in the way we are permitted to use our way of thinking.

Let me explain this better in a different way: If I said to you that there are very strange things happening as and within this Universe, would you be interested in knowing about them? I think your answer would probably be yes, tell me more about the strangeness as this Universe.

Yet if I told you that this God that exists as omnipresent is God-as-this-Universe-as-omnipresent, and that this God, that most of us believe in, is the same as what we have been scientifically calling Pure Energy, and that if as both of these titles, ***IT*** is the same being as what I have been referring to as ***IT***, that exists as something strange to our minds, as I have said before, it is because whatever this pure energy that is God exists in a way that could be construed as the reverse of how we exist, would you not think ***IT*** strange? Let me explain what I mean when I say that God exists in a way that could be construed as the reverse of how we exist: We exist as seeing things (matter) as being outside of us, for we cannot see our inner selves, because we were created to see ***IT*** more as being outside of us. And the reverse of this is that ***IT*** exists as a cold, clear, divine nothingness, that exists strangely as an inner and outer self simultaneously, and that ***IT*** is looking inside of ***ITSELF*** as ***IT*** reshapes what some call this empty Universe that exists as pure

energy, where God exists as omnipresent. I have had to adjust to the fact that I need to visually see ***IT*** outside of me in order to understand ***IT*** better, yet even before ***IT*** had our eyes to see, ***IT*** knew what to do with ***ITS*** clear, heated weight that also exists inside of ***ITS*** clear nothingness.

So now I have adjusted my understanding of seeing things (***ITS*** weight as matter) but understanding that ***ITS*** divineness and what we call or know as life, exists as being made of a cold, clear nothingness.

Knowing IT better

If ***IT*** allows us to understand ***IT*** better as ***IT*** exists as ***ITS*** dual energies, we might understand why ***ITS*** concentrated heated energy, may be equal to ***ITS*** cold, nothing energy. I say this because we know that we have grown tremendously from when we existed in caves to our present existence where there are more organized humans with minds that can see ***IT*** better, especially as the duality that pure energy (***IT***) exists as, and ***IT*** would be interesting to see if minds like that of Stephen Hawkins, and others can come together to use their intelligence and the information stored in their minds, combining all their information with all the information that now exists on pure energy (***IT***) in order to find ways to use the information that also exists stored in computers.

So, can we find a way to unite all the knowledge available about pure energy (***IT***)? I think so, because once we know that something can exist, it becomes easier to start a program that might be called The Center For The Investigation of ***ITS*** Knowledge, where as a group we could start finding ways to bring all the knowledge that exists on pure energy (***IT***), that is available in the now living human minds, and all the information that exists outside of humans, such as books and the enormous amount of information stored in computers and on the Internet. At this Center, all this information could be available for anyone to read, spawning the creation of a Third Internet, for there is already a 2 internet. This is a non-profit U.S.-based consortium of 206 universities and dozens of companies, including Sun Microsystems and Microsoft that are working together to improve the current Internet. Conversely this

group could generate a network where all the now existing information could be present for all to read as their findings. Furthermore, I am very grateful to the many minds, articles, and programs for information that existed on ***IT*** so that I could offer you, the readers, what little I have been able to see concerning pure energy (***IT***) and the way pure energy behaves, as ***IT*** reshapes ***ITSELF***. However, I again have to pause to avoid giving more importance to how ***IT*** exists as ***ITS*** omnipresence, and remember that this is not more important than remembering that ***IT*** already exists inside of me, and that ***IT*** already has shown me how to connect with ***IT*** through the 4 techniques that I learned from Maharaji, which surely are the same techniques that others may have already found, from other perfect masters, that were here before him. Furthermore, I am grateful that Maharaji's 4 techniques are very simple, so much so that anyone can use them to come in contact with ***IT*** in meditation. So, again I have to say: Thank you ***IT***, for sending a fragment of YOURSELF as Maharaji, to teach those of us that are interested in connecting with you as this moment of your existence.

To continue then on the subject of ***ITS*** weight as ***ITS*** weight reshapes, try to understand the following related statements: We know that things are changing very fast and we cannot seem to keep up with this very fast changing speed that ***IT*** is producing around us, but we should remember that we are in the midst of a technological revolution at this existing moment, and the same way ***IT*** was hard for people that lived during the era of The Industrial Revolution, it is just as hard for us to adapt to an ever changing society, now it is our turn to undergo a revolution which will force us to change, even if we still do not know how to adjust, for this techno revolution runs at a very high speed. I personally have learned to accept that what is happening out there is just a way that ***IT*** uses to reshape ***ITS*** weight into other possibilities, so that I do not fear what is happening, for everything out there is still totally ***IT*** as ***IT*** reshapes ***ITS*** weight. Even if things out there seem to be spinning out of control or moving at high speeds, I have noticed that in the very midst of it all I can enter the opposite state through being with ***IT*** in meditation, where I can still be with ***IT*** as just one

constant existing moment. Of course, as this constant existing moment change is still taking place, but inside of me, there is a peaceful moment that ***IT*** offers us as a place where we can be with ***IT***, without the effects of the turmoil that exists outside.

Nothingness as distance

Here is one more way to understand that ***IT***, as, has no **distance**, in the sense of distance to someone or something else. Allow me to clarify that statement: For ***IT*** (God) to have distance, there would have to be at least one more God, and then we could speak about the distance from one god to another. There is only one God and furthermore, we cannot apply distance to just ***ITS*** nothingness, for how can we measure something that is made of a nothingness?

⌘~~~~~~~~~~~~~~~~⌘ ⌘~~~~~~~~~~~~~~~~⌘

IT has no up or down, yet IT has an inside, which is where ITS weight exists. But IT does not have an outside, as ITS nothingness.

□~~~~~~~~~~~~~~~~□□~~~~~~~~~~~~~~~~□

ITS constancy

Here are some ideas that I will have to explain from my personal observation related to watching ***IT*** as ***IT*** exists and asking myself whether or not ***IT*** goes through what we go through, since we are made in ***ITS*** own image. The point I wanted to grasp was whether or not ***IT*** goes through changes the same way we experience the many changes that happen during our existence.

The only thing I found that made any sense is that whoever ***IT*** is, ***IT*** is beyond human understanding, or at least to me, for we should remember that ***IT*** has always existed, long before the Big Bang and will continue to exist, long after the stage we are in now.

As humans, we come and go, as do all the other life forms that now exist, but ***IT*** does not go through this, because anything that is alive is really just ***IT*** as one, and as ***ITS*** weight reshaping into all the existing possibilities, but ***IT*** is not affected by this, for ***IT*** still continues as ***ITS*** never ending SELF, since ***ITS*** transparent, clear way of existing does not change and has no ending.

This is easier to understand if we remember that this nothingness that ***IT*** exists as is the same nothingness that this Universe exists

as, and also, it is the same nothingness that exists inside of every existing thing, which is really just ***ITS*** less than 1% as ***ITS*** weight, which is what is changing, but not really being destroyed. That is why we say that pure energy cannot be created or destroyed, for how can we create or destroy ***IT***, as ***IT*** exists; as ***ITS*** weight that exists within ***ITS*** nothingness?

So, my conclusion is that it is ***ITS*** weight that is constantly changing, and this can only happen inside of ***ITSELF,*** as the clear, constant, transparent way that ***IT*** exists, that never does take part in change, or as having no beginning or end, for this would bring an end to how ***IT*** exists. But we know that this does not happen, because we now have enough information on ***IT*** as the pure energy that ***IT*** exists as, that we have been able to observe and understand some things about how ***IT*** behaves, and we know that ***IT*** does not have a beginning or an end as ***ITS*** nothingness.

⌘~~~~~~~~~~~~~~~~~⌘ ⌘~~~~~~~~~~~~~~~~~⌘

****There is a saying that" we cannot remove not even one atom from this Universe". Now this is so because, we cannot remove or destroy not even one fraction of ITS weight that exists within ITS nothingness**** *

❀~~~~~~~~~~~~~~~~~❀ ❀~~~~~~~~~~~~~~~~~❀

I am looking forward to seeing if we, as ***ITS*** weight, can manufacture equipment to understand more about how ***IT*** exists in this dual manner: as ***ITS*** cold, clear, transparent body where ***ITS*** less than 1% weight exists within ***ITSELF*** as ***ITS*** body. And now that we know that ***IT*** as this nothingness does exist, it will be easier to look for ways to understand ***IT*** in this manner of ***ITS*** existence.

But on the other hand, without seeming to be contradictory, this won't be easy because how do we make instruments from ***ITS*** weight that are already composed of 95% of ***ITS*** nothingness as the instrument itself, (because the instrument is made from atoms), that we will be using to detect something that is made of a clear, transparent nothingness? The one thing that we do have to go on is that this nothingness does exist as a freezing cold temperature, that ***IT*** presently exists as, as this Universe.

Scientists have found that this pure energy that is ***IT,*** exists as a form of nothingness that we see as the huge, cold, clear, transparent nothingness and which we call "The Universe", but this nothingness is also found as the tiny amount of ***ITS*** nothingness that exists inside every atom. By the way, the tiny amount of ***ITS*** nothingness that exists inside every atom would also be hard to measure. Nevertheless, ***ITS*** nothingness that exists as this Universe and ***ITS*** nothingness that can be found inside every atom are one and the same. The only thing that separates one from the other are the tiny fragments of ***ITS*** weight that we call electrons. If we had X-ray vision, so to speak, and could look at the Universe from a distance we would be able to see that there is only one nothingness, ***ITS*** nothingness, existing as this Universe and existing inside every atom, for all atoms exist within ***ITS*** nothingness.

So for now, all I know is that ***IT*** will continue to allow us to connect to ***ITS*** nothingness through meditation. But as to us acquiring more knowledge about how ***IT*** exists out there as ***ITS*** nothingness, we will have to see if, as this moment of ***ITS*** existence, ***IT*** will allow us to understand ***ITS*** constant nothingness through the invention of some kind of physical instrument or apparatus, for through meditation we already know that we as humans cannot understand ***ITS*** nothingness, but only experience ***IT***, since there is no way to document ***IT*** as ***ITS*** nothingness during meditation and we cannot use any instruments to know how this nothingness exists.

*** Help wanted. An engineer that can build an instrument that can explain or analyze a form of a clear, transparent nothingness***

⌘~~~~~~~~~~~~~~~~⌘ ⌘~~~~~~~~~~~~~~~~⌘

You can help

I end this section with a call to those who can see or have more information that relates to ***IT***. Please share it with us. I find it intriguing that ***IT***, as one, makes possible everything that exists as matter, heated weight. I know more information can be obtained as

more people focus on ***IT***. More scientific information specifically will lead us to understand not just the way ***IT*** exists out there--as Mars, Jupiter, or the Sun--but also as ***IT*** exists in all that can be found both within us as ***ITSELF***, and as heated weight within ***ITS*** cold, clear nothingness. This will not be so hard to see if we return to the photo at the beginning of this book and remember ***ITS*** duality, that just like the photo, ***IT*** is one with two extremes, as in the case of the photo of the old and young woman, and as the existence of ***ITS*** heated weight and cold clear nothingness.

Spin and matter
If you look closely you will see that the speed of everything that rotates is governed by the amount of weight that is attached to it. For instance, an electron will spin near the speed of 186,000 mps because an electron has very little of ***ITS*** weight attached. On Earth, much more of ***ITS*** weight is in one place, so our planet rotates at its present speed.
You too have these ingredients. You do not seem to be spinning because you are rotating in harmony with the spin of Earth. You do not spin as you walk due to your individual weight and because every electron in each one of your atoms is spinning.

⌘~~~~~~~~~~~~~~~~⌘ ⌘~~~~~~~~~~~~~~~~⌘

**** Men experience this spin when they urinate (women too, of course, but men typically have a better view). The spinning causes everything to move in the direction of the spinning at the border of our galaxy. ****

❀~~~~~~❀

Speed as motion
Here is a way to understand why we are in constant motion. Imagine that you are somewhere along the Equator and that you are sitting still. You might believe that you are not moving, but the truth is that you are moving at 1000 miles an hour, or 6 miles a minute, or 0.26 miles a second, which is 137.28 feet per second because of the rotation of our planet. And this motion continues beyond our planet because our planet is in a galaxy called the Milky Way by the 3 pound blob of water that calls itself the human brain--and this galaxy is 100,000 light years across--also determined by the 3 pound blob-and the galaxy is spinning at the

incredible speed of 400 miles per second. All of this spinning makes sense because anything that spins is transferring energy as ***IT***.

The water blob called the brain is also measuring this spin from the position where the brain finds itself. Since the brain is here on Earth, it is using the rotation of our planet to measure the Universe in terms of time and distance. I hope that the human mind in processing more up to date information that exists as this moment of who we are and why we are here will now better understand itself as a water blob and that it is measuring that which put ***IT*** together in the first place, as a place called omnipresent. The mind put the word "omnipresent" into existence, but can the human mind accept this word's meaning? ***ITS*** fragmented portions as ***ITS*** oneness are one, for there are not two of ***IT***. As one, ***IT*** exists in a dual form, as in ***ITS*** weight, which we know about as protons, neutrons, and electrons, and ***ITS*** nothingness (empty space), as the duality, or opposite of ***ITS*** weight. These are two opposites that can exist independently, as in ***ITS*** stage one (before the Big Bang).

So, in using ***ITS*** oneness as the electron to encircle ***ITS*** quantifiable weight within a given portion of ***ITS*** nothingness, ***IT*** could become what we now know exists as matter (atoms).

Matter (atoms) is where we can understand ***IT*** as all ***ITS*** qualities, but in fragmented portions: first, as ***ITS*** weight (protons, neutrons, and electrons); second, as ***ITS*** nothingness; and third, as both ***ITS*** weight and nothingness. And it is necessary for ***IT*** to exist in this way if we are to be alive. In this stage of ***ITS*** reshaping, ***IT*** used ***ITS*** quantifiable portions of ***ITS*** weight to produce protons, electrons, and neutrons, and ***IT*** placed these quantifiable portions of ***ITS*** weight within a quantifiable portion of ***ITS*** nothingness, as the empty space that exists inside the atom, and then ***IT*** held it there with the way ***IT*** can also exist as ***ITS*** oneness, as when ***IT*** exists as both weight and nothingness, which are the qualities of the electron. The electron may be a way to see ***IT*** as fragments of ***ITS*** oneness.

As we view an electron with a special high powered microscope, we see a portion or image of how ***IT*** exists having dual forces: as high speed nothingness and as the weight that also exists in the atom. But looking more closely, we will also see a third portion: first, the fragment of ***ITS*** weight (protons and neutrons), second, the fragment of ***ITS*** nothingness (the empty space in every atom), and third, that which holds these fragments within the fragmented portion of ***ITS*** total oneness, which is both ***ITS*** weight and nothingness as the electron.

And this portion of ***ITS*** oneness that is the electron should not be confused with the way ***IT*** could exist in stage three, which is where ***IT*** might exist as a total oneness, where ***ITS*** weight and ***ITS*** nothingness exist as one total blend, or form. But by looking inside the atom during this stage two of ***ITS*** existence, it is the electron that shows us the only real evidence of ***ITS*** weight, or a portion of it anyway, for I do not yet know scientifically if we could visually see or detect the nothingness that also exists there.

Then again, maybe ***IT*** will allow the development of instruments that will let us understand ***ITS*** way of existing as nothingness, as an energy force that exists inside and outside the atom as this Universe. And this will happen because if we were able to understand ***IT*** better, then ***IT*** too would understand ***ITSELF*** better, at least in words, since before and many rotations after ***IT*** reshaped into this planet and then us, there was no written information on ***IT***. But again, I feel that ***IT*** does not really need to exist in words, for ***IT*** knows what it is doing with ***ITSELF*** as ***ITSELF***.

⌘~~~~~~~~~~~~~~~~⌘⌘~~~~~~~~~~~~~~~~⌘

******Anything that has motion has this motion within ITS nothingness, yet ITS nothingness does not have motion within someplace else, for if IT did, this someplace else would still have to be within ITSELF.*** ***

❀~~~~~~~~~~~~~~~❀~~~~~~~~~~~~~~❀~~~~~~~~~~~~~❀

The Universe as 186,000 mps

Returning to our look at the Universe through the analogy of the atom, we see that matter cannot form at the speed of 186,000 mps, and since the Universe is basically empty of matter, it's likely the Universe as nothingness could be just the speed of 186,000 mps, which is to say that ***IT*** exists as nothingness at the speed of 186,000 mps or faster. After all, speed is an intricate part of ***ITSELF***, the same way that weight is an intricate part of ***IT***. This would explain how ***IT*** reshaped into matter: ***IT* reduced *ITS* speed**. In this way, in order to exist as matter, ***IT*** reshaped into electrons, which is ***IT*** as weight.

The Big Bang and 186,000 mps

The speed of 186,000 mps is scientifically known to exist independently in the absence of matter. As a result of the Big Bang, ***ITS*** weight (heated dense matter) was pushed outwards and slowed down when it encountered the speed of 186,000 mps: This is where electrons were born. This gave rise to what we see as matter. It's also what we see as sunlight being pushed outward, light behaving as both a wave and as particles; or photons catching a ride so to speak at the speed of 186,000 mps.

In order to have light or matter (atoms), ***IT*** has to be present as weight, and this weight has to be kept separate so that the different elements can be formed. To hold in this weight, ***IT*** uses ***ITSELF*** at a slower speed; this creates the electron. The same happens in the formation protons and electrons. These two types of particles, having opposite electrical charges, they repel each other.

In light, we have 186,000 mps, and we also have ***IT*** as a duality of ***ITSELF***. Just before the Big Bang, ***IT*** had all ***ITS*** weight in a place that existed in that part of ***ITSELF*** where the empty space still exists. If the speed of 186,000 mps automatically exists in the absence of matter, then when ***IT*** had all its weight in one place and exploded in the Big Bang, the scattered pieces, as weight, are what we now have as all the atoms that exist as matter. This makes sense because if ***IT*** existed as the empty space and the speed of 186,000

mps, then ***IT*** had all the speed that was necessary to form and maintain what we see as matter.

To me, everything that exists is ***IT***, and there is nothing we can do to change the way ***IT*** is. Everything I say and do now is a way of trying to better understand ***IT*** and the nature of our existence.

MAXX-SPEED AS black holes

It is said that not even light can escape black holes when they pull in the surrounding matter. Let us use this to understand ***MAXX-SPEED.*** When light travels at 186 thousand miles per second, it is carrying energy that can behave as a particle or a wave. If we could remove this particle and wave and just stay with the speed, we would then have light's maximum speed. I thus use the term "***MAXX-SPEED***" in place of "the speed of light" because of this and because “the speed of light” is a human phrasc that describes something that only came into existence after the Big Bang.

Let me return to the concept that states not even light can escape the gravitational pull of a black hole. I think there is a possibility: The pull of a black hole may be faster than the speed of 186,000 miles per second, otherwise light would escape it. Remember that light as speed carries energy as the weight that has to exist as that particle or wave, which ***IT*** produced after the Big Bang.
The pulling force is the way ***IT*** exists as that very cold temperature in the 99% nothingness. We know scientifically that cold will pull heat, and this is also why this ***MAXX-SPEED*** is able to exist in this 99% nothingness that ***IT*** exists as.

Where IT exists as speed

I have tried to understand how much ***MAXX-SPEED*** there could be as a total of ***ITSELF***, and have come to accept that ***IT*** simply exists naturally as this speed, and that ***ITS*** consciousnesses also exists in this nothingness or as this speed, as an energy in ***ITSELF,*** as the bulk of ***ITSELF***, which is more than enough for ***IT*** to exist as this consciousness and as the living things that exist. I have to add again that all the billions of humans that exist on this planet, with all the googols of life forms that also exist, all are ***IT*** as one

entirety of ***ITSELF*** as a consciousness, for everything that exists in this Universe is ***IT*** as one. And ***IT*** is one in large part precisely due to ***MAXX-SPEED:***

1. ***MAXX-SPEED*** acts as a carrier of ***ITS*** own weight, as a transferor of the energy that is carried outwards from our Sun, as weight, so that we could exist.

2. As ***MAXX-SPEED*** slows, ***IT*** can reshape into electrons, for which we should be thankful; since without electrons, we and everything that is made of matter could not exist.

We see this speed in music, for sound requires vibrations that move at certain speeds, and we see it too in cars, trains, and airplanes and not only the speed as miles per hour but as all of the speed that their shells or bodies have, which includes the speed of the electrons in every atom of their constitution. We can add here too that our sex lives involve speed: the speed of performance, the speed of sperm, and best yet, if we did not exist as the speed of every electron in our body, we couldn't reproduce. The sperm and ovary, made up mostly of water, are also governed by the speed of electrons. As you can see, this high speed is everywhere, even if this speed is not made of something that we visually see, such as matter.

How much MAXX-SPEED is IT?

As to how much of ***ITS*** nothingness exists at ***MAXX-SPEED***, it could be 1% or ***IT*** could be 100 %, which would include all Dark Matter since Dark Matter, it is believed, is not made up of protons, neutrons, and electrons, that would slow speed down. The only numbers that we can now add together as a close total would have to come from adding all the high speeds that exist in every atom.

Beginning with hydrogen, we count one high speed electron. And continuing through the elements, each being heavier, we would count more of ***ITS*** high speed electrons. If we could add all the electrons of the existing atoms in all elements of matter, we would get numbers that could be in the millions and trillions or more in

terms of speed. With this we would see that there is not just one ***MAXX-SPEED*** but many. And all of this nothingness that ***IT*** exists as could be made up of this ***MAXX-SPEED***.

MAXX-SPEED in a circle

MAXX-SPEED exists in a circular motion. At the moment of the Big Bang, ***IT*** threw all quantifiable fragments of ***ITS*** weight outwards into ***ITS*** already existing total nothingness. Let us call this quantifiable weight "number one", and we will see this even better if we refer to this one as the weight of a hydrogen atom's proton. For each proton weight, ***IT*** adds another independent portion of ***ITSELF***, an even smaller fragment of ***ITS*** weight, an electron. The weight in ***MAXX-SPEED*** that now exists as the electron behaves as an opposite force to the weight that exists as the proton. This in turn keeps the electron spinning inwards in search of its opposite, which happens in a circular motion.
The proton, having mass (heated weight) is a positive force, and the electron having significantly less mass is the negative force. As opposites, the heated weight pulls the speeded, cold nothingness that exists as the electron. But they will resist coming together because of the huge nothingness between them.

And when ***IT*** sent out ***ITS*** next heavier quantifiable heated weight, ***IT*** sent out more than one electron to keep this weight intact in the nothingness. This next heavier heated weight is what we have come to know as element number two, helium. And when ***IT*** added more quantifiable heated weight, ***IT*** became what we call element three, and as ***IT*** added more quantifiable heated weight ***IT*** became what we now call element four, and this goes on to the 92 elements found in nature (and also to elements numbers 94 through 118 that are not) that exist as ***ITS*** heated weight, which exists because of ***ITS*** circling high speed. To this we also have to add the nothingness that exists between the speed and weight of every atom. If it were not for ***ITS*** speeded weights and nothingness, we could not exist as the matter that we are composed of.

MAXX-SPEED as a straight line

IT also uses ***MAXX-SPEED*** in a straight line such as seen when our Sun projects light outwards. ***IT*** does this by sending out a miniature quantity of ***ITS*** weight which can behave as a wave or as a particle, which together catch a ride in an area that is made of nothingness. And as nothingness, the weight from our Sun will have no resistance to slow it down. There is nothing pulling it in a given direction. This high speed is, as one would say, free to travel until what we call light hits more of ***ITS*** existing weight, where ***IT*** will then deposit it onto the already existing heated weight of atoms, and being made of those atoms, I must thank ***IT*** for sending this heated weight as solar energy so that I can exist.

Furthermore, not only is the Sun pushing ***ITS*** weight as waves and particles, but this ***MAXX-SPEED*** also already exists as a cold nothingness, which is the ideal carrier. It is a scientific fact that cold pulls heat. So this ***MAXX-SPEED*** is eager to pull this heated weight and assist with the Sun's outward push. The heat moves straight outward in all directions where this ***MAXX-SPEED*** will find no resistance; there is nothing in nothingness to slow it down.

The ***MAXX-SPEED*** that carries this heated weight will deposit it as energy where it meets with matter (more heat). Maybe heat can only exist with more heat, as when this heat existed in a singular area, such as in the very dense matter that existed as ***IT*** just before the Big Bang.

The next time you feel the heat from our Sun remember that it is the same heat that ***IT*** exists as. ***IT*** sent it from our Sun so that we could continue existing, at least here on Earth. I am not sure how ***IT*** will energize us as we leave this planet, with our new clothing called space suits. We will have to wait for that information until the astronauts have documented it.

Dark Matter as MAXX-SPEED

Being a form of nothingness, Dark Matter will continue to elude our senses. We will never be able to physically feel it as speed because Dark Matter is not made of matter as weight, and this

MAXX-SPEED can only exist where there is no weight slowing the ***MAXX-SPEED*** down. Since ***MAXX-SPEED*** has no weight attached to it, there cannot be an impact with it, as in bumping into ***ITSELF***.

Speed as a builder

We can, however, thank ***MAXX-SPEED*** for matter. The next time you look at a building, like the Empire State Building, or the highest building in your area, or even a bridge, remember that after the Big Bang ***MAXX-SPEED*** became the high speed of the electrons in every atom of every element of matter. Buildings exist because of the atoms that compose the materials that make buildings. Materials such as steel are composed of atoms of mostly iron and carbon having weight in their interior (protons), which exists in a volume that is at least 95% nothingness. And at the outer part of this atom, you will find the electrons, which are basically speed with a little weight.

The speed that exists as electrons functions to hold on to other atoms in one of two types of bonding; ionic bonding or covalent bonding. It is because of this bonding of one speed with another, as electrons, that something that can stand on its own. It is the high speed of the electrons that hold things together with such a force that a building can go up as high as the Empire State Building, and higher.

When you look at a steel column that is only 5 feet tall, here again you are seeing ***ITS*** weight distributed as quantifiable fragments within 95% nothingness. The only power holding up the column as ***ITS*** weight is ***ITS*** speed. If you can see this, then you can also see that you can only stand or sit upright because your weight is distributed in fragments that exist in an area that is 95% nothingness, and all of this fragmented nothingness is held together by the speed of the electrons.

Speed and push

It is ***ITS*** speeded nothingness that is giving ***ITS*** stationary weight motion, starting as spin, such as the spin of electrons and planets,

and ***IT*** is as this spin, in the form of electrons, protons, and neutrons that we also exist.

Just a thought: Now, ***IT*** could be that ***IT*** uses ***ITS*** speeded nothingness to produce motion...Look at ***IT*** this way: Let's start at the moment of the Big Bang and start with ***ITS*** very dense weight that at that moment was existing as one stationary heated weight, as a unit, which was so heavy, that for ***IT*** to be able to move this one unit within ***ITSELF*** would have required a huge amount of energy to be able to push this one huge very dense weight, so what ***IT*** did was this: ***IT*** first fragmented ***ITS*** heated weight in to smaller pieces, so as to make this weight smaller and easier to move, and then ***IT*** used ***ITS*** speeded nothingness to: 1) push this weight outwards, and 2) ***IT*** then gave spin to this fragmented weight. So then ***ITS*** fragmented weight moved outwards, also as spin with the electrons, and ***ITS*** nothingness as speed repeated the same process to form protons and neutrons and also provided them with spin.

To summarize then, ***IT*** used ***ITS*** speeded nothingness to initiate a push outwards on ***ITS*** singular weight at the moment of the Big Bang, and as ***IT*** fragmented this singular weight into atoms, ***IT*** gave this weight motion, along with the motion that exists in the celestial bodies that are still until this day moving about within ***ITS*** nothingness, which we call the Universe.

MAXX-SPEED as a mover and carrier

MAXX-SPEED exists as the speed of light, which is 186 thousand miles per second, or any speed faster than 186,000 mps. We know that this ***MAXX-SPEED*** can only exist in the absence of matter (weight), so this ***MAXX-SPEED*** can exist in nothingness.

We can see the effects of ***MAXX-SPEED*** in action as ***IT*** carries weight as a particle (photon) from the ***Sun. This dual by-product is being pushed outwards, catching a ride on the MAXX-SPEED*** that exists in the empty nothingness and clinging to the speed in all directions. What makes this situation easy is that one of the

qualities of cold is that it pulls heat, so the heat that exists in the particle/wave of light already has an outward motion.

MAXX-SPEED exists in a no-weight zone as the nothingness of the Universe. Inside the hydrogen atom there is the speed of the electron, which is slower than ***MAXX-SPEED*** because of its weight. Its speed does not result from a process similar to the outward movement of sunlight. If you look at the hydrogen atom from the outside, you will see that the electron's speed results from the maximum speed that has always existed as the cold nothingness of ***IT***.

IT, for only ***IT*** knows how to do this, placed a very small amount of ***ITS*** weight, on this already existing ***MAXX-SPEED***, so as to fold in and rotate around the proton. Let's take this in reverse to see it better. If the weight of the electron **were** removed, this ***MAXX-SPEED*** would exist right where it is located, as the same ***MAXX-SPEED*** that exists in cold nothingness. Yet, we cannot see this nothingness; one, because ***IT*** is transparent, and two because it does not contain matter. So ironically, ***MAXX-SPEED*** must be made of nothingness in order to exist.

As for the possibilities of a contracting Universe, this can also happen within the expanding Universe. This would be as a result of this outer weight losing its outward push from the speed that ***IT*** has because the outer weight cannot escape from the total cold nothingness. We should remember that cold pulls heat, so heated weight will always be pulled in, or kept in, as having to stay within the total cold nothingness that ***IT*** exists as. Thus, the weight that is moving outwards will always have to stay within the cold nothingness because cold pulls heat.

This is an easy situation for ***IT*** because ***IT*** is 99.99% cold nothingness and less than 1% heated weight. And in this cold nothingness ***MAXX-SPEED*** exists, which is what ***IT*** uses to carry and move ***ITS*** weight; like what happens with the Sun when ***IT*** uses speed to maintain its weight within a given area, like what ***IT*** does in atoms within ***ITSELF***.

Regarding how much time ***MAXX-SPEED*** takes or uses, the first thing to remember is that ***MAXX-SPEED*** cannot be measured in relation to time. First, we need matter in order to see our human mechanical time system as something that changes. Second, we know that at the speed of 186,000 mps, our human time system does not apply; matter cannot exist at the speed of light.

So, knowing this you will understand that since ***IT*** is 99.99% cold nothingness, where ***IT*** also exists as this ***MAXX-SPEED,*** we cannot attach our clock to ***IT*** in order to understand ***IT*** in the three different ways that we have already covered; which were: one, this huge vast nothingness that we know as the existing Universe; two, as the nothingness that exists in every atom; and three, as the nothingness that exists in meditation.

As to ***MAXX-SPEED*** being independent, this speed could exist without a direction, for ***IT*** does not have an up or down. If this speed does exist, how does it know which way to go? I can see that this ***MAXX-SPEED*** could exist among itself, for there has to be more than one, otherwise we would only get one ray of light at a time. That isn't happening as far as the number of beams we receive as light. There could be a number that we have absolutely no way of accommodating into a word.

We will use our Sun as an example of why this speed is independent of direction. Speed needs a direction. And it gets one from the solar weight being pushed away from the Sun in every direction it can manage. It then travels in a straight line until it hits something. This is where it can drop the energy it carries as ***ITS*** heated weight. An example is when we get hit by sunrays; we feel the heat being transferred onto our skin. The heat comes from radiant energy transferred to us in order to maintain the conservation of pure energy. ***MAXX-SPEED***, which has no matter, also must obey the law of the conservation of pure energy. It could still be there where it hit me as a sunray, and is waiting for the weight of ***IT*** to push it in another direction. Another guess would be that the ***MAXX-SPEED*** returned to its weightless area, being that it is not governed by the force of gravity.

Speed and weight

Here are some ideas related to speed, but let me start by saying that on our planet the bigger an object is in terms of size and weight, the harder it is to get it to move faster without applying more and more energy. One reason why this is so, is because this object is made of ***ITS*** heated weight that has a built in characteristic that has been there since the moment before the Big Bang, when all ***ITS*** weight existed as one singular weight. Now, this characteristic of ***ITS*** weight wanting to become one stationary point as weight will always be there, and this is why we have to apply energy in order to move ***ITS*** heated weight, and whenever we try to apply speed to ***ITS*** weight, ***IT*** involves a complex mechanism. Let me give you an example: We humans, as ***ITS*** weight, can move ***ITS*** weight as our human bodies. But whatever speed we can achieve in moving our bodies is only possible because we are a complex system that can move ***ITS*** weight within certain speed limitations. Another example would be a jet airplane which is also composed of ***ITS*** heated weight. In order to move a jet airplane's weight, which is really again ***ITS*** heated weight as the atoms that make up the elements in the metal alloys that the jet is made from, this jet has to exist also as a complex system, that can take in energy in order to move its weight. And in order to move this jet plane's weight, other complex systems have to exist: its fuel system, its ignition system, its turbines, etc. But this jet and any other object, including us, that is made from ***ITS*** heated weight (atoms) will also have limitations as to how fast it can move.

⌘~~~~~~⌘

*** ***We use ITS heated weight as food for our fuel in order to move other things that exist as ITS weight.*** ***

❀~~~~~~~~~~~~~~~~~~❀❀~~~~~~~~~~~~~~~~~~❀

Now, I have been saying the above because for us to obtain faster speeds the way we now exist, we would have to leave behind as much of ***ITS*** weight as possible. That is why we say that nothing made of ***ITS*** weight can go faster than the speed of light.

Now we know why we cannot travel faster than 186,000 miles per second, because for us to push ***ITS*** weight to higher speeds, we have to leave as much of ***ITS*** weight behind as possible. ***IT*** will be easier to understand this if you go back to the picture of the two

ladies in the Preface, and remember that the importance of the photo is not the images contained in the photo itself, but the fact that both images co-exist as one photograph, there are two extreme opposite views that exist in one and the same picture.

So that if we put aside the viewpoint that we have as ***ITS*** weight and the limitations that this imposes on the speed we can achieve in movement and go to the other viewpoint, which is that ***IT*** already exists at a speed greater than 186,000 miles per second (which I call ***MAXX-SPEED***), you will be able to understand that this ***MAXX-SPEED*** is possible because ***IT*** has none of ***ITS*** heated weight attached to ***IT***, and that this ***MAXX-SPEED*** exists as the cold, clear nothingness that ***IT*** exists as, which we call the empty Universe.

In order to see that this ***MAXX-SPEED*** does exist out there as ***ITS*** nothingness you have to use your imagination: So imagine that as soon as this ***MAXX-SPEED*** slows down slightly, just by using a tiny amount of ***ITS*** heated weight, an electron is created, which is the point where all matter begins. However, beyond the electron as a particle of matter, we are back into ***ITS*** nothingness, where ***ITS*** weight also exists as the many particles that exist that are heavier than the electron, but not as heavy as protons or neutrons. But all these in-between particles of ***ITS*** weight (that we know as sub-atomic particles) still exist within ***ITS*** one total nothingness.

We as electrical speed

We can only be here to communicate about speed because ***IT*** reshaped into water, which gave us our existence and mobility. This is obvious because if we were made of steel, we could not move. Water is made of atoms that exist because of the speed that exists as electrons. Speed continues in our existence, giving our body parts, our hands and our feet, mobility by way of electrical impulses--electrical speed messengers--sent by the brain. This speed, which is slower than ***ITS*** maximum speed, is what, as electricity (which is a very high speed that is derived from the electron), gives our mind the ability to think and to send electrical impulses.

⌘~~~~~~~~~~~~~~~~~~~~~~~~~~~~~~~~~~⌘

******Without speed we cannot move.******

❀~~~~~~~~~~~~~~~~~~~~~~~~~~~~~~❀

We can see this speed as electricity, as the messenger from our head to our toes. We can see this speed in the electrical implants that have been installed in people who have lost motion in their arms, or legs, and we have seen it also facilitating the beating of the heart in pacemakers. So we now know that basically our brains communicate and function through electrical impulses and that our whole body is composed of this slower than ***MAXX-SPEED***. Therefore the only reason we exist is because ***IT*** exists as speed and because ***IT*** slowed down from ***ITS MAXX-SPEED*** by adding a fraction of ***ITS*** heated weight to this speed and reshaping into electrons, so that ***IT*** could then hold in more of ***ITS*** weight in the form of protons and neutrons, so that atoms could exist and we could then exist. We can only function because of this speed that we exist as, as the electrical body that we have, and we can only function, in terms of thinking and moving because of what ***IT*** exists as, as a divine, conscious speed.

One more reason to be grateful for ITS weight as electricity
Here is something interesting, about ***ITS*** oneness, concerning ***ITS*** existence as electricity.

First let me say, that if ***IT*** did not exist as electricity, or stopped being ***ITSELF*** as electricity, our developed nations would cease to exist, for we would not have light, and we would not get water which depends on electric pumps, our stores would not operate, and our trains and planes would not operate. We would not have electricity to pump our gas, no more communications, no TVs or radios, no electric stoves or refrigerators, for us to prepare and conserve our food. And without electricity we would not be permitted to leave this planet. These are just a few things that are ruled by ***ITS*** way of existing as ***ITS*** oneness (electricity), as the weight that exists as the particles that ride in this high speeded nothingness, called electrons, not to mention that if there were no electricity, all the electronically stored information in computers would be lost.

IT is speed; IT is divine; IT is conscious

I may not be an expert in this area of the many ways that speed exists in this Universe, but as readers may begin to focus on this, for what we have until now been focusing on is the weight of this pure energy, we may come into more information on the opposite of this pure energy and the nothingness that this ***MAXX-SPEED*** exists as. We can start by taking what we know as ***ITS*** weight, and turning it around, or upside down, so that we can see its opposites, for ***IT*** is there also, as the pure energy that we know exists, and one thing for certain is that we have grown mentally in information since we first arrived on this planet.

If there are readers who see more in connection to this ***MAXX-SPEED*** and want to share it with the rest of us, please send it to me so that I can post it in my webpage, which will display this incoming information as an open databank that can be viewed by all. As bizarre as all this information may be to the human mind, my mind has to be thankful that ***IT*** does exist as this electrical speed so that I can exist.

Electricity as ITS weight and speed

IT exists as a minute form of ***ITS*** weight in electricity: as the weight in the electrons that make electricity possible and as the nothingness that these high speed electrons exists as. In electricity we can see ***ITS*** duality as the high speed nothingness that exists as electrons and as the stationary, heated weight riding on this high speed nothingness. We can also see what happens when these two collide such as when a light bulb emits light and heat as ***ITS*** duality that continues to exist as light, which behaves as both a particle(***ITS*** heated weight)and a wave (***ITS*** speed, a form of nothingness).

The word electricity comes from the word electron, which is ***IT*** as a high speed form of nothingness. The more weight the electron has the slower it will move.

So, again thank you (*IT*) for lighting and heating our existence as your weight and as your nothingness. And yes, I have learned not to touch you as electricity.

Nothingness as speed

Here is an idea that relates to ***ITS*** high speed as the Universe: Since ***MAXX-SPEED*** can exist in this nothingness that ***IT*** exists as, I have questioned why this speed does not crash into ***ITSELF***, since ***IT*** is going in all directions, and the answer could be that since this speed is made up of a nothingness, there is nothing to crash into and it is when light picks up a tiny fragment of ***IT***S weight as light, that light exists. Now this does not mean that if we remove this fragment that light has as ***ITS*** weight, this ***MAXX-SPEED*** would not exist, for it is scientifically known that this speed can exist in the absence of matter. Here is one more way to understand this: If light is something, which it is, for we can see what light looks like as the brightness that we can see, (I say this because if light is a nothingness, as in not having weight attached, then the nothingness that now exists as this Universe, would also be lit up with light, but since light behaves as both a wave and a particle simultaneously, and particles must have weight. Now, if this weight is removed, then the speed of light could be faster, which is what I call ***MAXX-SPEED.***

Where is IT, in the Universe that IT starts, ends, and returns?

First we should remember that ***IT*** is ***ITS*** weight that has a starting point, which is that "place" when ***ITS*** less than 1% weight existed as one singular point, and it is this weight that ends up as black holes, so the entire circle may be completed, and all of this is happening within ***ITS*** nothingness.

ITS dual way of existing as weight

Now that we have reviewed what we know concerning ***ITS*** weight, you can use your mind in an alternative way, in order to see the many ways ***IT*** also exists as ***ITS*** nothingness.

Ionic and covalent bonding are instances in which ***IT*** uses a tiny fraction of ***ITS*** weight as electrons, so as to lock in ***ITS*** other

fragmented weight as protons, in order to hold ***ITSELF*** together as different possibilities as matter, for instance, as when matter is flexible, as in certain metals, or as being rigid (stiff) brittle, as in concrete.

Covalent or chemical bonding is one more way to see what ***IT*** does with ***ITS*** tiny amount of weight as electrons. In this case ***IT*** can share one atom's electrons with an other atom, making something even stronger.

Let me explain this better this way: ***IT*** can take ***ITS*** weight and hold ***IT*** together rigidly by using ***ITS*** minute weight as electrons to hold atoms together very strongly, for instance, in the case of concrete, forming types of matter which are strong but brittle, while in other cases ***IT*** can use the tiny weight of electrons to form flexible types of matter, like the steel rebar's, in a concrete building, that can bend without breaking.

Will the Universe end?

Scientifically, we know that pure energy exists as a place that has no beginning or end. As pure energy, nothing is ever created or destroyed. Nothingness cannot be destroyed for there is nothing to destroy. And if it could be destroyed, where would ***IT*** exist? The Universe is made from ***ITS*** nothingness.

What we call destroying is simply a rearrangement of ***ITS*** weight. To say for instance "the Universe can collapse", is to say that there is a whole lot of something that can fall into itself, as it were. But nothingness cannot fall into nothing, can it? Matter, weight, can collapse into itself, but not nothingness. And mostly, nothingness is all there is.

We may try to destroy it, but to no avail. Our efforts can only cause transmutation, the changing or reforming of something into another thing, but we cannot actually destroy anything into total annihilation or oblivion, in the sense of totally ceasing to exist.

There is an immutable law: The conservation of pure energy causes transmutation. It does not matter what we do; things are transmuting, not being destroyed. The shape and matter of something may be destroyed, but the pure energy that it was still is, but somewhere else, doing something else. Our insistence in seeing things destroyed is our mind thinking primitively, believing only what we see and perceive. This is natural, but it keeps us from seeing ***IT*** as ***IT*** is, something that simply exists, and luckily, as something that cannot be destroyed.

IT just continually reshapes, thank you very much! And here we are.

I am grateful to science for helping me understand ***IT*** better. Through science we understand pure energy and many things that go with it. The technology that makes it possible to understand the technical part of ***IT*** makes it easy to understand how we did not understand ***IT*** before.

I will continue to be grateful to all those scientific minds. And I do mean all, because biology shows us how ***IT*** reshaped into humans, geology shows us the reshaping into our planet. And so on. But I must remind the scientific minds that physics is simply the study of how ***IT*** behaves as pure energy.

So much scientific information is now available. I have tried to see it all, crazy me! One good source is the internet, obviously. A good website is **www.newton.dep.anl.gov/askasci/phys98.htm**.

When I queried: "Why does pure energy have no beginning or end?" My result was that "physics does not offer an answer to the question 'why,' but one of the most solidly based observations is that energy is conserved; that is, it can be transferred from one form to another, but neither created nor destroyed. Physics does not deal with what energy is; it deals with how energy behaves." Of course, I don't see ***IT*** scientifically, that is, through the eyes of a scientist, but from the point of view of a human being that is constantly reshaping. No college degrees are necessary for us to

understand ***IT*** better. And for those readers who are interested in sharing their findings related to pure energy, I've provided contact information at the end of this book. Your insights are also welcome as postings on the databank of my webpage: **http://www.ricricardo.com**

Here is another way to see things that are accepted as scientific information. We know that pure energy cannot be created or destroyed because this energy has always existed. We cannot remove one atom's worth of energy. This pure energy is what makes up our planet and every atom that makes our bodies possible. And since everything that exists is this pure energy which has no beginning or end, all ***IT*** is doing is transmuting or reshaping.

When we refer to something as having become extinct, what we really mean is that this pure energy has reshaped into some thing else. All extinct species, like the dinosaurs, have simply transmuted.

The Universe as 100% IT

We cannot unify the energies that exist as pure energy (God) because they already exist as one.

The most I can see and understand as ***IT*** being unified is the idea of the Universe being made of 100% emptiness, and that within this 100% emptiness ***ITS*** weight and mass may exist in an infinitesimally even and undetectable distribution.

⌘~~~~~~~~~~~~~~⌘~~~~~~~~~~~~~~⌘

The Universe does not have a beginning or end as ITS nothingness **

❀~~~~~~~~~~~~~~❀~~~~~~~~~~~~~~❀~~~~~~~~~~~~~~❀

As big and as small

Try to see this Universe as a whole, as one, and then narro*w* your scope to see some of the smaller areas within ***IT***. Look at galaxies, for example. We can see that they are made of energy that remains in the form of galaxies, and reducing our view to an even smaller area of ***IT***, such as planets, we continue to see that they too retain their form as planets, and when we see ***IT*** reduced to a smaller area while still being within ***ITSELF*** as the total Universe, we can see

humans, each being a unit of ***ITSELF*** existing within ***ITSELF***. From here, we can see a reduction of ***IT*** as very small cells and bacteria, and we can continue to reduce until we see ***IT***, which began as big as the Universe, as a single atom. But even here, ***IT*** can reduce ***ITSELF*** to the many events that take place within the atom.

With our human capability of understanding, we recognize these infinite activities in this area so small within the atom, that we have considered these activities to be a Universe within ***ITSELF***, and as a duality, the atom is a miniature Universe within a huge Universe. This leads to speculation that there are yet other Universes, and I would agree, but to start to see those, we will have to put the word "mini" in front of the word "Universe."

⌘~~~~~~~~~~~~~~~~~⌘⌘~~~~~~~~~~~~~~~~~⌘

** ***There is no open or closed Universe, for IT does not open or close as ITS nothingness*** ***

❀~~~~~~~~~~~~~~~~~❀❀~~~~~~~~~~~~~~~~~❀

No parallel Universe

There cannot be a parallel Universe for various reasons:

1-The word Universe means that there is just one Universe.

2-Whatever this other place is, it would also have to exist as ***IT***, as omnipresent.

3-For this other place to exist, it would have to be made of something, which means that this other place has to have ***ITS*** weight also, and if this other place could exist, it would also have to exist within ***ITS*** 99% nothingness.

So, this other place, like any other place that could exist, still has to exist as the one hundred percent that ***IT*** exists as.

I find that my mind is fascinated that it has found ***IT*** interesting, for I now know that anything that can exist has to be ***IT*** as God, or as pure energy.

Now I try not to get lost with the many things that we see and understand that are made from atoms (***ITS*** weight) that must exist

within ***ITS*** nothingness as that place which we refer to as this Universe.

If there were another Universe that is parallel, where would this place exist? The next time you look at all the matter that is distributed throughout ***ITS*** nothingness, you will also notice that all of this matter also has within itself, as matter (atoms), the empty nothingness that exists in every atom, and this nothingness in every atom can only exist within ***ITS*** total nothingness, not in any other place parallel to ***ITSELF***. So, for there to be a parallel Universe, it would have to be using the same total nothingness that ***IT*** exists as, for this other parallel Universe cannot exist outside of ***ITSELF*** because this would mean that there is more than one God or more than one pure energy in action.

⌘~~~~~~~~~~~~~~~~~~~~~~~~~~~~~~~~⌘

*****IT does not exist parallel to anything.*****

❁~~~~~~~~~~~~~~~~~~~~~~~~❁

For similar reasons I do not believe that if a black hole exploded there would be a new Universe. To begin with, all the black holes that have been detected already exist within what we refer to as the Universe. Furthermore, as the word indicates, there is only one **UNI-VERSE**, because if there could be another Universe, we would be indicating that ***IT*** has more than one body and more than one heated weight, which would cause one to arrive at the untenable conclusion that there is more than one GOD, and more than one pure energy in existence.

The word Universe

The word Universe may need some redefinition. Webster's dictionary defines Universe as a noun listing the defining terms as cosmos, creation, the visible world, astral system, universal frame, all created things, everything, nature, the natural world.

I would further define the given terms as follows:

1.Universe: ***ITS*** housing or shell as God, as pure energy; how God, as pure energy, exists as ***ITS*** heated weight and ***ITS*** cold nothingness; the huge in size and extension of ***IT,*** which exists as a form of cold, transparent nothingness.

2.Cosmos: the result of God's reshaping ***ITS*** heated weight and speed throughout the Universe.
3.Creation: the effect from Gods' reshaping ***ITS*** heated, dense weight in to quantifiable fragmented weight (atoms = matter) in what we have come to see and understand as having happened after the Big Bang.
4.The visible world: God's weight and nothingness reshaped into something for all eyes to see.
5.Astral System: the way God moves ***ITS*** heated weight as energy around ***ITS*** cold, clear nothingness.
6.Universal frame: God's shell as a housing that exists or as composed of 99.99 % cold, clear nothingness; that which holds God's heated weight.
7.All created things: God's ability to permit us to participate in ***ITS*** reshaping.
8.Everything: nature, the natural world; a place where we can exist as a moment of ***ITS*** existence.

Now that you have read these definitions you will see that what we have been referring to in the dictionary is all referring to ***IT***, as ***ITS*** weight (matter). And we account for ***ITS*** nothingness as the hugeness of this Universe when we refer to traveling from one planet to another, all of which exists as this Universe. We know that this Universe is that part of ***ITSELF*** that exists as the nothingness, or as that which we refer to as Dark Matter, as a word, for we know that this Dark Matter is not really made from matter as in atoms (***ITS*** weight).

If we ever find a better definition for the Universe, it would have to also include ***ITS*** 99% nothingness that this Universe exists as, rather than solely what we have up until now been referring to as the matter that exists as this Universe or within this Universe, which when totaled equals less than one percent of ***ITS*** entire extension. As I have mentioned elsewhere, what we refer to as this Universe is really ***ITS*** shell or outer housing, something like the housing we have as our outside body that is holding in everything that exists inside of us as weight.

Remember the photo you saw at the beginning of this book, of the young and old lady. In the same way that ***IT*** is just one totality, the photo is just one photo, yet it has two different images; now look at ***IT*** the same way you did the photo, remembering that the photo is one and try to see both extreme images simultaneously. See the Universe the same way. We are already attuned to seeing it one way, as the heated matter in action such as what the planets and stars do as ***ITS*** weight. Now look for ***ITS*** other extreme image, ***ITS*** 99% high speeded, freezing cold, clear nothingness.

⌘~~~~~~~~~~~~~~~~~⌘⌘~~~~~~~~~~~~~~~~~⌘

****** The Universe is where it exists as ITS nothingness, where ITS weight exists. ******

❁~~~~~~~~~~~~~~~❁~~~~~~~~~~~~~~~❁~~~~~~~~~~~~~~❁

We may be at the center of the Big Bang

It may be that we are in the middle of the Universe and also in the center of the Big Bang, and this is why planets and galaxies appear to be moving away from us.

The Universe is God

If we humans could create, then there would be no need for a God. Hence we have the expression, "God created the Universe." God has been introduced to us as something that exists “up there, in Heaven”, in a place He created, and along with Heaven, he created the Earth and the whole Universe. But this implies that God is something apart from Heaven, Earth, and the Universe. If God created the Universe, for example, ***IT*** would first would have had to find a place to put this Universe, and ***IT*** would then have had to find the materials necessary to make the Universe. And ***IT*** would also have had to put all the pure energy of the Universe into the Universe as though it had previously been someplace else or non-existent. But ***IT*** couldn't have gone and found this pure energy; all that could exist already existed and still exists. Nor was pure energy non-existent before the formation of our Universe because ***IT*** cannot be created or destroyed. The scientific community validates this.

When we were taught about God, we had very limited information as to how God exists as omnipresent or as this Universe. But in my

observations of the Universe and its contents, I know that all is God as one that is omnipresent. ***IT*** will always be the same ***IT*** that I am talking to and the same ***IT*** that listens; there are not two Gods, just as there are no two pure energies. Scientifically, we know that the Universe cannot exist if it were not for the pure energy (God) that the Universe is made of. Knowing this makes it clear: The Universe is God, and God is the Universe. Neither God nor ***IT*** created or is creating the Universe; ***IT*** reshapes it as ***IT*** transmutes ***ITS*** heated weight, so as to continue to evolve as ***ITSELF***.

I have no problem using the word God when I refer to the pure energy or the Universe, for we know scientifically that pure energy exists and is the key to our existence. I want you to remember certain things that will help you to understand this universal and omnipresent ***GOD***. When science began to develop, we knew that something existed out there that could not be created or destroyed, and we called ***IT*** pure energy. Now we have the God that exists for us as humans, and we have science's God that exists as pure energy. But regarding the God that exists as pure energy, some scientists neglect to include our planet, our moon, or all the galaxies; these things, including us, as also being this pure energy. Luckily, it makes no difference to ***IT*** whether or not we remember that we and everything that exists as matter, ***ITS*** heated weight, are also ***IT***. But now that we know that ***IT*** exists as 99.99% cold, clear nothingness, as the omnipresence that exists as this Universe, we can better understand God as ***IT*** and ***ITS*** omnipresent nothingness.

Furthermore, when we view the Universe as something apart from God, we focus on celestial matter, which is natural, for our mind needs to see something in order to accept that something exists. Our mind should remember that even though it was conditioned to only accept what it can see, matter is less than 1% heated weight that exists inside cold, clear nothingness. It is a fact that this nothingness exists and that this nothingness is pure energy and ***IT***. So the next time you look at the Milky Way, remind yourself that the stars and celestial bodies are only 1% of ***IT***. They are a part of ***ITS*** weight, and the rest of ***IT*** is nothingness. The mind should

also practice accepting that ***IT*** too is 1% heated weight within nothingness, the same nothingness that exists as the Universe and in every atom. This nothingness makes it possible for me to write about ***IT*** as the nothingness that is the pure energy or the God that we say created this Universe. For as this nothingness, ***IT,*** is conscious of ***ITS*** reshaping ***ITS*** weight; ***IT*** was conscious before the Big Bang and after.

IT has always reshaped ***ITS*** weight, but has never reshaped ***ITS*** nothingness. This makes sense because how can nothingness be reshaped? We have never seen the possibility of there being a Universe that had ***ITS*** weight on the outside, and ***ITS*** nothingness in the inside, and this also applies to matter, for matter exists because ***ITS*** weight, in the form of protons and neutrons, is inside, not vice versa.

IT is now moving ***ITS*** weight and nothingness into tinier and tinier fragments that we know exist as nanotechnology. But first ***IT*** reshaped into humans that had eyes to be able to see what ***IT*** was going to do, and hands to do these things as one more existing possibility of the things ***IT*** can do with ***ITS*** weight. That we are conscious and that we can think, feel, and the many other things that we do, are all forms of nothingness made possible because ***IT*** either has these attributes or ***IT*** knows how to reshape to produce them from nothingness, which is something as pure energy and the same pure energy from which we derive our consciousness. Our being conscious does not come from ***ITS*** weight, for we, as of yet, have been unable to weigh our thoughts, as in thinking, or weigh what we call being conscious. These attributes only exist as ***ITS*** nothingness, ***ITS*** conscious, omnipresent nothingness.

One more way

If ***IT*** reshaped into everything that exists on this planet as ***ITS*** weight, there could be the possibility that ***IT*** used ***ITS*** weight on some other part of ***ITSELF (the Universe)*** as ***ITS*** weight to reshape ***ITSELF*** with mobility, as life. I do not doubt the possibility that I could have inter-planetary cousins as the result of ***ITS*** reshaping. ***IT*** very well could use ***ITS*** weight within ***ITS***

nothingness to form extra-terrestrial bodies, especially when here on Earth ***IT*** used ***ITS*** weight and nothingness to produce very pretty humans in all colors and sizes, having minds that can know everything exists as ***ITS*** weight within a huge, cold nothingness (the Universe), which we now understand as the omnipresent ***IT***.

Future astronauts will see and share more information related to the things that ***IT*** exists as, as other parts of ***ITS*** weight and nothingness; and I hope that as these future astronauts come in contact with these other parts of ***IT,*** they remember the meaning of *omnipresent*, for to do so is to understand ***IT*** as God, or that which exists as pure energy. When we, with our human minds, can accept what this word implies, we, as minds, can accept ***ITS*** total omnipresence, and we, as minds, will then be able to answer the questions: **Who am I? and Why am I here**? To which my mind for now answers "Thank you ***IT***, for permitting me to be here as a bit of your weight surrounded by your divine, conscious nothingness."

You might better understand ***IT*** if you recall we are not who we think we are; we are ***IT*** as ***IT*** evolves.

⌘~~~~~~~~~~~~~~~~~~~~~~~~~~~~~~~~~~~⌘

*** ***Who are we?, we are IT as IT evolves*** ***

~~~~~~

No neighbors

The new frontier, is the place where you will not have a problem with too many neighbors or with overpopulation, where hardly anyone has moved in yet, but you will be living closer to ***IT*** for you will be living closer to ***ITS*** nothingness.

Why the Universe is not brighter

Keeping in mind that light is a by-product of our Sun fusing hydrogen atoms, we can see that for our Universe to have more light, there would have to be more stars. But this cannot happen because in order to produce more stars, there would have to be more atoms or the atomic weight that ***IT*** uses to form stars, but all the weight that exists is all that ever existed as ***IT***, so there cannot
~~~~~~

be more, not until ***IT*** reshapes our Sun to be more stars, for our Sun as ***ITS*** weight will also transmute in omnipresence.
As I have mentioned elsewhere, however, our planet is still going strong and will make millions and millions more rotations before our Sun will explode. So as our planet makes one more turn and another after that, we can continue to say that we have yet another day to use our human mechanical time system to transfer energy. Astronauts too have far to travel while ***IT*** prepares a new home and environment for future astronauts.

⌘~~~~~~~~~~~~~~~~⌘⌘~~~~~~~~~~~~~~~~⌘

Darkness is just the absence of a by-product called solar light.

❁~~~~~~~~~~~~~~~~~~~~~~~~~~~~~~~~~~~~❁

The atom as a model of the Universe

Some say the Universe is expanding, and I agree from the perspective that ***ITS*** movement is the result of the outward motion of the Big Bang explosion. This may also be the duality to the idea that the Universe will eventually collapse and destroy itself. Nonetheless, ***IT*** will not collapse to bring an end to the Universe. ***IT*** will bring all of ***ITS*** weight back to a central point again as if it were a reversal of the Big Bang. But the Universe is not expanding, what is happening is that ***IT*** is moving ***ITS*** weight around inside of ***ITS*** nothingness.

Let us review the Big Bang. ***IT*** was in the form of very hot dense matter, which had extreme and ultimate weight. This occasioned the Big Bang, in which ***IT*** spread this weight as heat throughout the Universe in the form of atoms, giving us all the elements. But let us not get hung up on that. Yes, we are talking about original matter exploding. This is where we get time, which cannot exist without motion. But the Universe is not merely matter in motion; it is about emptiness, a cold nothingness that is from 95%-99% of everything that exists. This truth applies to pure energy, matter, and everything that ***IT*** is constantly transmuting as.

Now let us try something: Instead of seeing the Big Bang exploding outwards, let us see it in reverse. Bring the Universe back to when ***IT*** was one ultimate heavy mass of weight, remembering that all we have done is compress matter, which is

heat as weight. We now see a very dense matter centered in the same nothingness that now exists as this Universe. And this would also look just like a proton inside a hydrogen atom. The proton exemplifies the matter -the less than 1% heat and weight of the Universe- kept in a place that is 95%-99% empty space. To hold its weight in place, however, the atom needs the aid of speed, which we know as the electron, which travels the perimeter of the atom near the speed of light.

⌘~~~~~~~~~~~~~~~~~~~~~~~~~~~~~~~~⌘

The speed of light can circle our planet ten times in one second.

❀~~~~~~~~~~~~~~~~~~~~~~~~~~~~~~~~❀

What you are seeing as an atom is the way ***IT*** was just before the Big Bang. But let's not overlook the emptiness of ***IT***. We talk about and analyze ***ITS*** infinite arrangements and reshaping, but while the weight of ***IT***, as objects, has a beginning and an end, what about ***ITS*** invisible shell of emptiness, ***ITS*** housing as a nothingness?

⌘~~~~~~⌘

*** ***The atom is a model of the Universe. And, the Universe is a model of the atom, for they are both IT.*** ***

❀~~~~~~~~~~~~~~~~❀❀~~~~~~~~~~~~~~~~❀

Pure energy as intelligence

There did come a moment when humans were the most intelligent form of ***IT*** on this planet other than ***ITSELF***. The progress of communications demonstrates the reshaping of intelligence into the high-tech form ***IT*** is today. At this stage in ***ITS*** progression, technology is out-performing the human mind in many areas. ***IT*** is now moving by space travel from this planet to other parts of ***ITSELF*** in the Universe. Spacecrafts utilize computers, which utilize microchips, all of which were not available before or shortly after the Big Bang.

This is all part of ***ITS*** plan and not something for you and me to question. And we still do not know how ***IT*** operates; we have only tried to describe ***IT*** with the small bits of information we have acquired from what we call the past. For example, we are now more aware of how pure energy works. We understand what ***IT*** takes to produce the effects of the Big Bang and that after ***IT***

reshaped from dense matter into a perfectly functioning Universe, the illusion of solid matter came. And it remains my belief that if we could weigh this matter, it would weigh the same as the dense matter that existed before the Big Bang, further demonstrating what we know about the conservation of pure energy.

IT, as energy, functions within ***ITSELF*** to keep everything that is happening within ***ITSELF***; nothing of ***IT*** is ever gained or lost, which is the same as saying, "pure energy cannot be created or destroyed." For pure energy is ***IT*** as a total. And as a total and always being the total, ***IT*** was never created or creating; ***IT*** has just been reshaping. We can see an example of this reshaping when we imagine ***IT*** as the very dense matter that reshaped ***ITSELF*** into the existing Universe. And we will see more events happening within ***IT*** when we accept that ***IT*** is the Universe, and all the energy transferred within this Universe stays in this Universe. The conservation of pure energy is therefore the same as ***IT*** transferring energy within ***ITSELF***.

Everything is in ITS image

After observing the things that exist in this whole Universe in regard to ***IT*** as omnipresent, I find that we, like the biggest ships or the tallest buildings, like the mountains, oceans, and seas, and even the Sun and Moon, are all made in ***ITS*** own image. Everything that is alive: the air you are breathing, the water you drink, the food you eat, and everything that is not alive, like the ships or satellites, and the bridges all have in common their existence in ***ITS*** own image.

I am referring here not to the way ***IT*** looks, but rather to what ***IT*** is made of, which is the same energy and nothingness that we exist as. The book you are reading, the TV and VCR, everything that you own as property is part of ***IT***. The stuff as matter is not what is important here. What is important is that all existing matter in this Universe is composed of atoms. Look at what makes matter possible, and see the similarities between those qualities and ***IT***. All matter is made of atoms. And all atoms have the following:

1.Weight, as in the weight of protons and neutrons.
2.Speed, as in the speed of electrons.
3.The nothingness that exists between the atom's nucleus and the electron or electrons.

All the above qualities are the same as those what ***IT*** exists as, and they cannot:
1.Exist without ***ITS*** very dense weight.
2.Exist without the ***MAXX-SPEED*** that ***IT*** also exists as.
3.Exist without the nothingness that ***IT*** also exists as.

So that if you look closely you will see that these are all the same qualities as an image of how ***IT*** exists.

Satellite transmissions

Some will disagree when I say that all our satellites and probes into outer space are also ***IT***. You will agree if you remember that everything we send out there is made of atoms, and atoms are made from pure energy. Even if we hold dearly to the idea that we created those satellites and probes and positioned them in space, we should remember that to make these satellites we had to use our brains, which are made of atoms as pure energy, we also had to use the materials to build them, and these materials are also made of atoms from pure energy.

As you can see, I have no problem with seeing everything as ***IT***, as pure energy.

How big is IT?

Big is only big to our human way of understanding size. Using an ant as an example will work if we overlook the problem that an ant cannot compare size and distance the way we do. But let us say, hypothetically, that the ant could cover the distance going around this planet. To the ant, our planet would be huge. It may even be possible that if the ant started this journey early in life, it might just see going around this planet as possible. Imagine now that the ant had to cover the distance spanned by our galaxy. Now, this situation would be impossible. This gives us a sense of ***ITS*** size.

As we travel further within ***ITS*** hugeness as our galaxy, the Milky Way, we will be moving on a road that is not firm; there will be no road at all, for this hugeness that ***IT*** exists as, to human understanding, is made of nothingness, and because of this, we will not be able to use our unaided human bodies, which are supported by skeletons made mostly of calcium, that exists on a firm ground (but not solid), and is held down because of gravity.

We will, with ***ITS*** help, reshape our present bodies, so we can live in an environment that does not have a firm ground. Nor will our new bodies be affected by gravity, at least not as we travel from one part of ***ITSELF*** as a planet to another if moving to a space where we are again governed by gravity. When this becomes our existence, we will be shaped into a new arrangement as a human body that can exist in ***ITS*** nothingness and move around ***ITS*** hugeness.

Humans who travel into ***ITS*** hugeness will have to do it without references such as up or down, beginning or ending, as to where they will be. If I were out there, I would have to say "Oh God, thank you for letting me see more of who You are."

So a word to the ones that will travel and see more of ***IT***: Always remember that if ***IT*** took you there, ***IT*** will protect you, for you are ***ITS*** pure energy in every atom you exist as. ***IT*** will help you to remember that when you say that you will be traveling into no man's land, you will actually be traveling into ***ITS*** nothingness. And never forget that as you move around ***ITS*** nothingness, you are protected by the most powerful energy that exists.

Astronauts will be the first ones that will experience the changes that ***IT*** needs to make so that we can travel into outer space in ***ITS*** nothingness. The way we are now is based on the organs that we have because we exist in an environment with gravity. Our hearts are made to pump blood upwards, and our waste is made to fall downwards. Astronauts already know about the problems with our human body in outer space.

No man's land

We, and where we find ourselves, are here because of matter. All that exists takes place in this area where matter already exists, and as matter, it will have a beginning and end. As we travel into the interior of ***IT,*** which we call the Universe, the first thing that happens is that we leave land behind, which means that we will no longer be using the phrase "keeping our feet on the ground."

We will have to begin to think in a different way because we will then be traveling in something empty, in nothingness, where there will be no up or down, and we will begin to become aware that ***IT*** as omnipresent is also this nothingness in which we are traveling. What will then give us a sense of distance will be the weight of ***IT*** as heat and matter because here is where we will have to get used to existing in a place where there is no beginning or end as something that is made of matter.

We will have to forget about the phrase "to the ends of the Universe," because there is no end. The Universe is not made only of matter that is less than 1% of ***ITS*** total; ***IT*** is also an infinite nothingness. And I know that as humans we can change our way of thinking when our environment changes.

⌘~~~~~~⌘

Any discoveries we make will bring us closer to understanding IT better

~~~~~~~~~~~~~~~~~~~~~~~~~~~~~~~~~~

Pure energy + [Plus]

To say that pure energy has no beginning or end is to indicate that pure energy is something that exists as having no beginning or end. Thus this something that exists as omnipresent has to be ***IT*** because one cannot exist without the other.

We already know pure energy has weight, and this weight is 1) visible as the fragmented matter that now exists in this Universe, and as the same singular weight that existed at the moment of the Big Bang; and 2) this weight is quantifiable, for we have been able to weigh it scientifically as the weight that exists in every atom and as the pure energy that we are composed of, as our human weight,
~~~~~~~~~~~~~~~~~~~~~~~~~~~~~~~~~~

for we personally see this weight every time we step onto a scale to weigh ourselves as the pure energy of which we are composed.

Now, adding the information we already have on ***IT*** as pure energy and 99% nothingness, we can more clearly understand that pure energy exists as a plus, as in a 99% form of a nothingness *plus* ***ITS*** weight that is less than 1% visible as matter. Let me explain as I do to friends when I tell them about pure energy: Pure energy is something (***IT***) that exists and is made of a plus + 99.99% form of energy that we can see as this cold, clear empty Universe which we refer to as Dark Matter, which is really composed of cold, clear nothingness. This pure energy that is less than 1% of what exists as heated matter (weight) is a minus that has to exist within this +99.99% nothingness that this pure energy exists as in ***ITS*** totality.

We can better understand this pure energy if we remember that what we will visually see, as this pure energy is what ***IT*** exists as in the less than 1% that exists as ***ITS*** weight throughout this +99.99% pure energy shell that exists as a form of nothingness which we cannot see or touch but that we know exists as being very cold and is what we refer to as the existing empty Universe.

Using the word *plus* to define this pure energy as a +99.99% form of a nothingness will take some adjusting to because this *plus* is really referring to something (***IT***) that exists as a cold, clear place (the empty Universe) that we understand as a shell of nothingness. We are accustomed to visually seeing ***ITS*** heated weight as our Sun, but we now need to understand that what we are seeing as ***ITS*** weight can only exist within ***ITS*** +99.99% cold, clear shell of nothingness. A little mind boggling?

Seeing a number like +99.99% in reference to nothingness may be a little confusing, but ***ITS*** expanse, which is the size of this existing empty Universe, is bigger than ***ITS*** weight, which is smaller in size, so ***IT*** must exist within ***ITSELF***. This was so even when ***ITS*** weight existed as a singular dense matter (weight) that was there just before the Big Bang. This weight was still smaller,

less than 1% smaller, when ***IT*** spread ***ITS*** less than 1% dense weight throughout ***ITS*** +99.99% clear, cold nothingness.

We can refer to ***ITS*** nothingness as being +99.99% because this is where ***IT*** is bigger in size, even if this number refers to nothingness. As to whether ***ITS*** less than 1% weight could be heavier than ***ITS*** 99.99% nothingness, I can only refer to what scientists say about one of ***ITS*** dualities, which is that ***IT*** exists in the same way an atom does, and as such, the proton (weight), or ***ITS*** positive part is equal in force to ***ITS*** negative (the near nothingness that the electron exists as), so it could be that ***ITS*** weight is equal to ***ITS*** nothingness as a force. We should remember that as ***ITS*** weight, protons can decompose into even smaller fragments, and this weight can behave as a wave, which is also a form of nothingness.

And as to what happens when ***ITS*** duality exists as one, which is when ***ITS*** weight and nothingness become one, we will have to hold off until ***IT*** finds a way to show us what happens at stage three, which is when ***IT*** has ***ITS*** less than 1% heated weight evenly distributed throughout ***ITS*** cold 99.99% nothingness.

Our numbers are minus

All that exists in this Universe is ***IT***. ***IT*** is 100%. When we say that there are five billion stars in a nearby galaxy, what we should remember is that these five billion stars are but a fraction of the total 100% of ***ITSELF***. Thus when we use numbers to count and measure, relative to the total 100%, our numbers are actually negative: less than 100%, less than one.

Now, coming back from the far end of this Universe, let's add into the picture our planet Earth, and forgive me some repetition here, but in viewing the information one way, you will see one thing, and in viewing the same information from a different direction, you will see another. Imagine that you are standing on the Moon and looking at Earth. We can see our planet because it has weight, as mass and color. The planet will look like a ball of solid matter

suspended in a dark empty substance, yet our planet, just as the atoms that comprise it, are 95% empty space.

If I could go to that area we call outer space; though I will never be able to, but if I could, I would take an electron and a proton and "wrap" the electron around the proton so that I would get what we know as an atom, and if I kept putting more of these atoms together, I would, with proper conditions, have a planet and every thing that we would find on a planet such as ours, including you and I.

This is why we are here on this planet, in the same empty space that exists as the one you see while being out there. ***IT*** reshaped ***ITSELF*** from the same space we know as outer space into what we know as this planet, while at the same time remaining the same empty space that it was. That which we see and say is out there, is not entirely out there, for we are in and part of that same empty space. As we stand in this empty space, we are standing in what we see as being out there--deep space or outer space.

We exist as omnipresence. Because we are made of matter and are separated from other blobs of matter however, we can say we are not in the same place as what we see out there. The difference is that in this place that we call Earth, ***IT*** reshaped to form an atmosphere that has oxygen where ***IT*** could exist as life. So, ***IT*** reshapes as atoms, such as oxygen, that are so crystal clear that we can see right through them.

I am grateful that ***IT*** has reshaped into this very clear atmosphere where we can exist comfortably because, as the astronauts have found out, it's very cold out there. And do not forget that we exist in a very, very tiny part of ***ITSELF***, so tiny that our whole galaxy is not even the size of a grain of sand with respect to the size of the Universe.

As I look around and see the things that ***IT*** has done, I am convinced that I have never known anything else that can do what ***IT*** does--such as taking an atom and reshaping it to where I now

find myself. I look at my body and see fingers that move because of the bone joints that exist as ***ITS*** duality, as a positive and negative, such as ball and socket joints, that exist as part of a perfectly functional skeleton that only ***IT*** knew how to shape, giving me a chance to exist with mobility. Even if I exist only as a human thought that can communicate with other human thoughts, I find this beyond incredible.

And remember, all that I have just mentioned is happening as billions of ***ITSELF*** in an empty substance that ***IT*** exists as. I will continue to repeat this because even if we think that matter is something solid, it is not.

We can see empty space in our surroundings, between ourselves and all outside us: our homes, work, play areas, and between the planets in our Solar System. We can see where there is light, which ***IT*** uses to reshape ***ITSELF***. We can see matter because ***IT,*** by making matter to absorb and reflect light and by giving us eyes and brains to see color, keeps us from seeing through all of ***IT***.

⌘~~~~~~⌘

Everything that IT reshapes into will always be 100% of the total

❀~~~~~~~~~~~~~~~~~~~~~~~~~~~~~~~~~~❀

Pure energy within

Please go back to the photo of the young and old lady and remember that what you are looking at is something that has two extremes. In this same way the features I mentioned above always exist within this one place that ***IT*** exists as.

Now you can look at pure energy from a different point of view. Instead of looking at everything outside you as being out there, try seeing ***IT*** as if you were inside this pure energy. We cannot exist outside of ***IT***, as pure energy as God, as a place that exists as ***IT***.

This will help you because you will begin to see things that are there. And just as with the photo, you might at first not see the other lady, you might not see the other side of ***IT***. Just say to yourself, "OK, I can see this one; now let me look for what is missing, as ***ITS*** duality." And remember that what is important is

that you will start training, exercising your gift of being able to think. And do not worry about what others may think, for you are not primarily here to serve them, but to transfer ***ITS*** energy.

What you see may not necessarily be of importance because no matter what you find, you cannot change ***IT***, and ***IT*** is the only one who is the 100% of everything and is running this place that exists as omnipresence.

You will be contributing to the way ***IT*** operates in whatever you may find. You will be participating in one of ***ITS*** dualities. ***IT*** will be looking for all existing possibilities, for that is ***ITS*** nature and that is why we have come into existence as one of these possibilities that ***IT*** exists as.

If you find this hard to believe, just remember that every atom that makes your existence possible is or comes from ***IT*** as pure energy.

IT as 100%

Another way to conceive of ***IT*** as being 100% of everything that exists is as follows: Think of ***IT*** as the pure energy that exists as the freezing cold, empty Universe along with everything that exists inside it which is matter. All this is made up of the pure energy that now exists as this empty Universe; both are just one, as in 100% something, and this something (***IT***) has no beginning or end, and it is inside this freezing cold Universe where this something has been moving ***ITS*** heated weight inside of ***ITSELF***, since long, long before we came into existence, and as one moment of this something's existence, ***IT*** used ***ITS*** heated weight in such a way to become you and I and every one else that exists as life as ***ITSELF***, because whatever this something is, it is the 100%, so that we, like every planet that may exist, is also this something as ***ITSELF*** as ***ITS*** 100%.

ITS neutrality

IT is one, having dual forces. We know this based on the way ***IT*** exists, or at least what we understand of ***ITS*** existence according to scientific information about pure energy. The Universe consists of

matter, which we refer to as ***ITS*** weight and as the protons (positively charged) and neutrons (neutral) that exist inside every atom. ***ITS*** weight, as matter, exists within the same empty nothingness of this Universe, which some refer to as Dark Matter. This Dark Matter is ***ITS*** negative force that makes up 99.99% of ***ITSELF***. Yet when we look at the Universe or what we see out there in space, we only account for the planets, stars, and anything that we can see, for our minds have not yet adjusted to understanding that nothingness also exists as a great part of ***ITS*** pure energy. I too used to overlook the nothingness, but now I am more aware of the emptiness inside every atom and the same empty nothingness of outer space. I have had to readjust my way of understanding this pure energy that exists and that I prefer to call ***IT***, and let me say that I still call ***IT*** *God*, for there are times when I have to call ***IT***, and it sounds better when I say "Oh God, thank you for letting me exist in your presence."

So, ***IT*** exists as weight, which is what gave us protons and neutrons as pure energy, and ***IT*** exists as weight's opposite: ***ITS*** nothingness. We have information that ***IT*** existed as these two forms even before the Big Bang. But we should remember that when ***IT*** existed before the Big Bang, ***IT*** was a singular concentrated weight (***ITS*** positive force), and a total nothingness, ***ITS*** negative force.

We have stacks of information as to what happened to ***ITS*** weight at, and after the moment of the Big Bang. But here we ignore ***ITS*** nothingness at the moment of the Big Bang when we saw, or referred to, ***ITS*** fragmented weight (matter) moving outwards. And we have kept viewing ***ITS*** weight as it moves within ***ITS*** nothingness (the now existing Universe), and we still overlook the fact that this weight, as matter, exists within ***ITS*** nothingness, and that this matter also consists of ***ITS*** total nothingness.

So far, we have no information to there being a neutral force present before the Big Bang. It is thanks to electrons that atoms can be neutral.

For me to understand ***IT*** better, as this now existing Universe, I have to see ***IT*** as the neutral or balanced way it exists, for ***IT*** was in this stage of ***ITS*** reshaping that it took a tiny amount of ***ITS*** speeded nothingness and a tiny amount of ***ITS*** weight and placed them together so that ***IT*** could reshape into what we now see as matter.

It is in viewing ***IT*** as this neutral or balanced way of existing that I can offer you what I am seeing.

As we look at matter as things or objects, what we are seeing first is ***ITS*** speeded nothingness. If I look at a person, or better yet, a steel wall, what I have to accept is that to see that steel wall, the first thing that has to exist to make that wall visible is ***ITS*** speed, as the outer speed of the electrons. In order for the steel wall to exist it has to be made of matter: atoms; and atoms can only exist because of the electrons that exist as the outer shell of the atom. Electrons can only exist because of ***ITS*** high speed form of nothingness.

Now imagine that for the steel wall to exist, atoms have to exist, and that atoms are what ***IT*** reshaped into using ***ITS*** dual qualities. The first is ***ITS*** 99.99% nothingness; the second is ***ITS*** less than 1% weight. If we could, which we cannot, see the way an atom exists, that is what we would see as far as what ***IT*** is doing with ***ITSELF*** as ***ITS*** speeded nothingness and weight.

Let's return to the hydrogen atom, the nucleus of which is made up of one quantifiable portion of ***ITS*** weight as the proton. As we head outwards from the nucleus, what exists, scientifically, is ***ITS*** nothingness, which as we move further out is cut off by its one electron. There you have the hydrogen atom: one proton, a huge nothingness, and then, existing as the outer surface of this nothingness, one electron, which has been considered by science as the opposite and negative portion of the atom.

Here is what I see, and not as a scientist, for I am nowhere near having the education that scientists have. But I am going to slightly

disagree with what is called the atom's negative being the electron. I say slightly disagree, for I too have to use this positive and negative language in order to communicate with you as it relates to ***IT***.

We refer to the proton as ***ITS*** positive, heated weight. Since we are using ***ITS*** heated weight as the positive, the opposite is said to be its electron but is actually ***ITS*** cold nothingness. We know this because ***IT*** existed before the Big Bang as very hot dense weight, not as the matter (atoms) that now exists. And this very dense weight existed within 99.99% cold nothingness. And what existed then has to exist now as the same pure energy because ***IT***, as pure energy, cannot be created or destroyed. Thus ***IT*** transformed. So let us go to the atom's heated positive weight that is the proton and move outwards.

As we leave this proton, as a positive heated weight, the first thing that we will contact is ***ITS*** negative nothingness, as the huge empty space that exists between the proton's heated weight and the spinning electron. Here is where we will use the information we have on ***ITS*** cold nothingness, and we will also use the rules that govern hot and cold, such as that cold pulls heat. The heat in the proton is being pulled by the cold nothingness in every atom. Since the proton, as heated weight, is the positive, then the opposite is ***ITS*** cold, negative nothingness, and since the proton is where the heat is, we should remember that as you get farther away from heat, temperature drops.

Also, ***ITS*** heated weight is in constant change, transmuting to give shape to other smaller things. In radioactive elements protons shed fragments of weight which are transmuted into radiation. In non-radioactive elements, protons and neutrons are made from smaller, sub-atomic particles. These sub-atomic particles, smaller than the proton, model what ***IT*** does as ***ITS*** weight inside ***ITS*** nothingness. The weight of all atoms in ***ITS*** huge, cold nothingness of the Universe are also where these even smaller sub-atomic particles, as weight, exist, for they both have to exist within ***ITS*** one total nothingness. There are not two different places as ***ITS*** nothingness,

and everything that has weight came from ***ITS*** original very dense, less than 1% weight that existed pre Big Bang.

IT pulls, as coldness, from the heated proton, weight to form the sub-atomic zoo that now exists within the atom. If this is true, then we will better understand ***IT*** this way:

As the atom, ***IT*** placed ***ITS*** heated weight, as the proton, within ***ITS*** one of a kind negative nothingness, and to prevent this heated weight from leaving a given area within ***ITS*** total nothingness, ***IT*** formed a neutral entity, which is what we now refer to as the negative electron. But in essence, this is a neutral force because the electron consists of ***ITS*** weight (the sub-atomic particles that make up the electron) and ***ITS*** speeded nothingness.

Now let me stop to mention something that I have mentioned before, as a personal point of view, that to me anything that exists as ***IT***, that exists in all three stages, is closer to the way ***IT*** really exists, for I can only see the electron in ***ITS*** stage two, which is post-Big Bang. In stage one, the term *positive* refers to the very dense mass just before the Big Bang. And it is easy to say that ***ITS*** mass exists as a positive because we know that it exists as something, for it has weight. Meanwhile, we label the electron as a negative, because it also exists, and we know this because it also has weight, as the particles that exist in the electron, but its weight is far less than that of the proton. And we jumped from the proton to the electron because we had no way of classifying the nothingness in between.

So if we analyze what we now understand as a positive, which is referring to ***ITS*** heated weight as something that we understand exists, the opposite is what we have not gotten accustomed to analyzing: nothingness. But when we analyze ***ITS*** existence as stage one and we apply our labeling to ***ITS*** heated weight as the positive, we should then also apply the word negative to ***ITS*** freezing cold nothingness as this pure energies opposite.

Furthermore, when we refer to what happened after the Big Bang, we only refer to how ***ITS*** weight reshaped into atoms, but we totally ignored ***ITS*** nothingness. The reason for this is clear; ***ITS*** nothingness is where ***IT*** exists as a constant. As we view the present Universe, what we see is ***ITS*** ever-changing weight, for if we look at what has been written, we will see that we are only stating that ***IT*** (pure energy) existed before the Big Bang as very dense matter, but very seldom refer to ***ITS*** nothingness, and we still write about ***ITS*** ever-changing weight as the Universe that now exists, and we still mention little of the nothingness that existed before the Big Bang, or the nothingness that still exists as this Universe, or the nothingness that exists inside the atom.

In stage one of ***ITS*** existence there are no electrons. So the difference in the stage that we are now in, stage two, is that ***IT*** is now using both ***ITS*** forces (speed as nothingness and ***ITS*** weight) to become a neutral energy (the electron) to divide ***ITSELF*** into even smaller quantifiable fragments of ***ITS*** total weight to form what now exist as atoms.

It is in this electron where ***IT*** becomes a neutral force, as an entity, which consists of ***ITS*** weight combined with ***ITS*** speeded nothingness. This is why I say that this combination is what would make the electron a neutral, for it is not just weight, like the proton, and it is not just empty nothingness; it consist of both a positive and negative, making ***IT*** more neutral, as a barrier or place holder, so to speak, between the negative nothingness that exists inside the atom and the negative that exists outside the atom.

When ***IT*** puts together a hydrogen atom, it keeps the fragmented weight as an entity in itself by separating it, as weight, within its nothingness with a neutral entity (the electron). Let me line up ten hydrogen atoms side by side. From left to right you see that ***IT*** placed a quantifiable amount of ***ITS*** weight, then gave this weight extension, then ***IT*** placed a neutral spinning force around this weight (the electron). To the right, ***IT*** again placed another quantifiable amount of ***ITS*** weight and did the same as the first atom. Between one fragmented weight and another, there is

distance, and the atoms are separated with the distance as ***ITS*** one total nothingness by another neutral electron. In using this neutral electron and a quantifiable portion of ***ITS*** heated weight, with a separation of this weight ***ITS*** as distance, ***IT*** could behave as an entity in itself.

If we could see a group of oxygen atoms next to each other, inside each we would see eight individual fragments (8 protons) neutralized from each other with a neutron. Better still, since energy of the same type (positive or negative) will repel, each proton is trying to push the other away, so by ***IT*** placing a neutron in between each proton, ***IT*** forces ***ITS*** own weight to stay together in a given area with in ***ITS*** nothingness. ***IT*** neutralizes the 8 fragmented protons that are inside the oxygen atom by applying one electron for every proton that can exist inside the atom; since oxygen has 8 protons, ***IT*** placed 8 neutral electrons to neutralize the force of the 8 protons.

So, ***IT*** uses ***ITSELF*** as both ***ITS*** forces (negative and positive) to form the electron which acts as a neutral barrier between the other fragmented weight (atoms), and ***IT***, as electrons, produces a separation between ***ITS*** negative nothingness.

So as I see it, the hydrogen atom is composed of a center that is where ***IT*** places ***ITS*** quantifiable heated weight within ***ITS*** cold nothingness, and ***IT*** then surrounds this weight by placing ***ITSELF*** in a dual form (+ -) as the electron, which ***IT*** uses as a neutralizing barrier so that we can exist as ***ITS*** heated weight and cold nothingness.

You might see it better this way: All electrons are on the outside of ***ITS*** negative nothingness to hold in ***ITS*** positive fragmented weight so that the different elements can exist. As always, we should remember that however we see or understand ***IT***, it will not change the way ***IT*** is. Luckily, our existence cannot alter the way ***IT*** was, is, or will be as ***IT*** continues ***ITS*** reshaping into what we will see on our way to understanding ***IT*** as the past, present, and future.

In this stage two of the Universe, ***IT*** is using both positive and negative forces. The electron, as ***ITS*** neutral way of being, makes it possible for ***IT*** to reshape into you and I and every thing that is out there as matter (***ITS*** weight) in the Universe. You will see this better when you remember that in order to have matter we need atoms, which consist of ***ITS*** weight (positive force) as protons and the 95 % or more nothingness that exists inside the atom, surrounded by ***ITS*** dual forces of weight and speeded nothingness that ***IT*** reshaped into as a neutral electron.

So, the next time you get in a car, or train, or plane, or boat, and go onto the roads for the car, and the track for the train, and into the air for the plane, and onto the water for the boat, remember these can only exist because of the neutralizing electron. Be grateful that ***IT*** exists as a neutral force, for this neutrality holds together the vehicles as the electrons that exist in every atom. This applies to your existence too, for you are also made from ***ITS*** positive force, as all the protons that you exist as, and as all the negative emptiness inside every atom, and the neutral force (the electron) that holds ***ITS*** fragmented weight within ***ITSELF***, as the nothingness that ***IT*** exists as every atom of matter, without which we would be back to ***ITS*** stage one.

The names we have given IT

For the longest time we have been registering ***ITS*** parts and the distance between them. We know that ***IT*** exists and we have given names to the closest parts to us, like the planets Earth, Jupiter, Venus, Mars, etc. As we have gone out farther into ***IT*** with our telescopes we have given ***IT*** names like the Milky Way, Andromeda, and others. The list is even bigger than these few which I have mentioned, for we now have many books on the names of the stars, nebulae, galaxies, and the distances between them that we know ***IT*** exists as. You can find these under the subject titles of astronomy and astrophysics.

Galaxies and black holes

I recently read an article in a magazine called Science News that said, "Two massive black holes are spiraling towards each other in

a gravitational dance that will end in a few hundred million years, when the black holes merge." To understand my explanation of the concept quoted, apply the knowledge that this whole Universe is ***IT***, that which signifies our God and that which is understood as pure energy.

Remember too that this pure energy was once a very dense matter and that ***IT*** existed as this dense matter. The pure energy then reshaped into atoms that are made up of 95% empty space, which enabled ***IT*** to reshape into this Universe.

The importance of black holes is that they are a manifestation of ***IT***. From them we can understand how ***IT*** compresses into dense matter. This matter is estimated to weigh tons per teaspoon. In the process of taking the matter in a galaxy or galaxies and compressing it, ***IT*** is functioning in ***ITS*** own image. ***IT*** reshapes itself as atoms that are 95 percent empty space; compressing ***ITSELF*** back into that very dense matter that existed prior to the Big Bang.

As far as the idea that this "gravitational dance will end in a few hundred million years," there are a few things we need to know:
·These activities are taking place in a place called omnipresent.
·When we talk about time, as in years, we are referencing the rotation of our planet. These activities are taking place with materials that exist in a no-time zone.

Time is the weight of ***IT*** spinning in the Universe. This Universe is the same Universe that existed before the Big Bang as 99% empty nothingness. What is changing is the use of ***ITS*** weight as ***IT*** reshapes. Better still to our understanding is that this 99% nothingness is not anything, for it is ***IT***, as something, and the most I can see is that this nothingness has expansion as a freezing cold temperature, and exist as an opposite temperature to ***ITS*** heated weight as the two pure energies that exist in this place called omnipresent.

And this is also because of the weight of ***IT*** as matter that exists from matter to matter as distance, so that at least we have something to go on in terms of measuring ***ITS*** weight as matter to matter as a way of measuring its size. We have to remember that distance is not something physical; it is a way of thinking. I have had to adapt to thinking of distance in a place unlike Earth that has no beginning or end.

Standing at a distance from the scenario of the black holes that the above article says will merge millions of years hence, if we count the 365 times that one million rotations our planet will have to make, we will see this process taking place not as a moment in time but rather as ***IT*** reshaping ***ITSELF*** as black holes that exist within ***ITSELF*** as space.

If we removed our planet or our galaxy or ourselves we could no longer use our concept of time. Our planet would not be there to give us a yearly rotation, and the human mind could not stamp these events as being millions of years away. Even without the prediction of years and our human concepts of time and space, the black hole merging process described in the article would still take place. It would not happen as a moment but as a place where ***IT*** has always existed and has only been reshaping ***ITS*** weight.

What came before "the" Big Bang? More big bangs?

In order to best understand what follows, let us first start with our understanding of the Big Bang. According to modern science this was a moment when all the Universe's heated weight was concentrated into a single point called a singularity, which for reasons unknown, exploded. As a result of the explosion all this heated weight became fragmented to become atoms, that now compose the 4.6% of the Universe that we know as matter. As this heated weight was thrown outwards, ***IT*** acquired motion, from one of the ways ***IT*** also exists, which I have called ***MAXX SPEED,*** Since then, ***IT*** has been existing as a duality: 1) as ***ITS*** freezing cold shell body which is none other than the freezing cold, empty nothingness that extends throughout the Universe, and 2) as ***ITS*** heated weight that forms all the matter that now exists inside this

freezing cold nothingness as this pure energy.

It is clear that ***IT*** threw ***ITS*** heated weight outwards into fragments within ***ITSELF*** in order to explore other possibilities that ***IT*** could exist as. I am very thankful for this because otherwise, this galaxy we exist in, along with our particular solar system with all the other planets inside, and very especially, this one planet what we call Earth would not exist as ***ITS*** heated weight, which means that you and I would not be here either, using ***ITS*** energy, in the form of our human body and ***ITS*** heated weight in the form of our brains. Neither would this book you are reading exist as ***ITS*** heated weight! For this reason I say thank you, ***IT***, for allowing us to exist in this manner, so that we could come to acknowledge your existence and understand a little about the qualities of your existence, in terms of your dual temperatures.

Returning to what ***IT*** is doing with ***ITS*** heated weight that ***IT*** threw outwards within ***ITSELF*** at the moment we call the Big Bang, I would like to summarize the information that scientific minds have gathered throughout human history concerning the existing moment called omnipresent, ***IT*** is logical to infer from observation that, in the same way ***IT*** began the Universe by exploding ***ITS*** heated weight outwards into fragments, ***IT*** will and is at this very moment accomplishing the process of bringing back, so to speak, ***ITS*** heated weight to one singular compressed point within ***ITSELF*** once again, through the collapsing process of supergiant stars or novae becoming black holes.

Imagination is a wonderful tool, so I invite you to imagine that you have huge arms and are somewhere in "outer space" grabbing every bit of matter that exists, let's say, in any given galaxy, piece by piece. The way I visualize ***IT*** is using my left "hand" to do this and then "squeezing" the contents to eliminate the 95-99% empty space that exists inside what I have collected as ***ITS*** weight, in terms of matter. This imaginative description is precisely what a supermassive black hole accomplishes. Then, I would repeat the entire process with my "right" hand, so as to end up with a supermassive black hole in each "hand". I would repeat this

process with each and every galaxy in the Universe, and every bit of stray matter. Finally, I would compress these two supermassive black holes into one, in my "left" hand. This would result in all ***ITS*** heated weight being compressed into a single, clear hot spot (please refer to the piece where I discuss the nature and properties of a clear hot spot) with an immense gravitational pull.

The process I have described above in an imaginative way is the reverse of what happened during the event that we call the Big Bang. What I have described is the process by which ***IT*** is reuniting all ***ITS*** heated weight that exists in a given galaxy and repeating this process until all ***ITS*** heated weight which is now dispersed throughout the Universe is brought back into a single point once again. This is accomplished by removing the emptiness that exists in matter and the use of the force of gravity for compression.

In fact, in this process all ***IT*** is doing is removing ***ITS*** weight that exists in the form of electrons in each atom that forms matter, so that the emptiness that exists inside each atom can reunite with the emptiness that exists outside each atom as the one total nothingness that exists as ***ITS*** shell body.

In addition to the above, another way of visualizing this process is that ***IT*** is not really removing the emptiness that exists inside each atom, but all ***IT*** is doing is removing the tiny amount of ***ITSELF*** that ***IT*** placed outside each atom in the form of electrons, in order to reunite ***ITS*** outside nothingness with ***ITS*** inside nothingness, for all ***ITS*** nothingness is just really ***ITS*** one total shell body. In this way, after ***IT*** removes ***ITS*** weight that exists in the form of electrons, ***ITS*** weight that exists inside each atom in the form of protons and neutrons will be united with the electrons to become one single heated weight. This is what black holes do.

After this process what will be left is a multitude of black holes or clear hot spots, scattered throughout ***ITS*** freezing cold body, (or what we call the empty Universe). What will happen next is that

these scattered hot spots, or black holes that possess an enormous gravitational pull, as they come close enough to each other, they will pull each other in, one black hole "eating up" the other, becoming larger and denser, and thereby augmenting ***ITS*** gravitational pull. This will cause the process to accelerate, until all the clear heated hot spots consume each other until all that remains will be one transparent hot spot, where all ***ITS*** heated weight will find itself merged into a singularity within ***ITS*** freezing cold body. You will surely realize that this is the same situation that existed before the Big Bang that produced the Universe we now know. I consider ***IT*** a logical deduction that at this point ***IT*** would "ignite", so to speak, another big bang explosion, so that this singular heated dense weight will be thrown outwards back into ***ITS*** freezing cold, clear body once again, so as to reshape ***ITSELF*** into a different possibility, so that ***IT*** can exist as a new universe within the way ***IT*** exists as a freezing cold, constant nothingness, where ***IT*** exists as omnipresent as being, alive as a divine consciousness.

I would like to insert a personal note here, for I have to be grateful to ***IT***, because ***IT*** was one of these "new" big bang explosions as ***ITS*** heated weight that ultimately gave rise to the conditions that allowed us to come into existence, which likewise, has allowed us to understand ***IT*** better in the terms that I have discussed in this piece above as well as in what follows, for every time ***IT*** becomes a new big bang explosion, which would be a new universe as ***ITS*** heated weight only, perhaps with added ingredients that ***IT*** knows about from the previous ways into which ***IT*** has reshaped ***ITS*** weight. As an example, when ***IT*** reshaped ***ITS*** weight into atoms and then into matter, and then into galaxies with stars and planets, where ***IT*** reshaped ***ITS*** weight into water, so that ***IT*** could have the necessary mobility for the existence of life and therefore, living human bodies, that gave ***IT*** the possibility of endowing these with consciousness and intelligence, beginning as cave dwellers that are ultimately to become space travelers, who have gone as far as discovering nanotechnology.

So, what will ***IT*** reshape into when this new reshaping occurs? Only ***IT*** will see this, because ***IT*** will be ***ITS*** weight that will be

reshaping into other, new possibilities, within ***ITS*** same, constant, freezing cold, transparent body. Whether ***IT*** chooses to reshape again into humans or into something beyond a human form is really up to ***IT***. It is fascinating to speculate about what ***IT*** may reshape into as ***ITS*** heated weight. ***IT*** could be a form of the way ***IT*** exists as life that may be totally different from how we now exist on this planet now. As things stand now, as space travelers we will have to readjust to existing within ***ITS*** nothingness, where the force of gravity is practically non-existent. Surely, we humans, who began life on in a warm atmosphere, leave this planet and start reproducing in ***ITS*** freezing coldness, where our solar system's Sun will not be the same stars. For all we know, whatever star may be in the vicinity may not even radiate strongly enough to even warm our skin. This will not be all! While traveling inside ***ITS*** nothingness, the humans of the future will encounter timelessness, because they will have left behind what we have been referring to as a 24 hour day, and when they leave our solar system they will not be able to use the rotation of our moon to measure months or our Sun to measure years. These future humans will be living in a place that is in itself timeless because ***IT*** is composed of a freezing cold nothingness that exists as omnipresent, where time cannot be measured, for we can only apply the concept of time to ***ITS*** heated weight as ***IT*** is in motion within ***ITS*** clear, transparent, freezing cold, empty nothingness.

Who knows what type of skin these new humans will have? Even their internal organs and body systems may change! Just consider how far we have come since we were cave dwellers. Nowadays there are people who have metal screws that hold parts of their skeletons together and plastic pumps that help their heart beat!

Such thoughts lead me to gratitude. So thank you (***IT***), for allowing me to understand you, in terms of your heated weight, which is my form of existence, and how you exist as life, and divineness and consciousness, for these are some of the qualities that I exist as, only because you exist as these qualities. This surely sounds strange and I fully realize that I may never really understand all of you, as you exist as your heated weight that exists

within your freezing cold shell body.

Now I would like to add a few more things concerning to ***ITS*** existence in terms of dual temperatures, because I feel that there might be some readers who may connect this to their knowledge of thermal energy. If so, and you would like to share ***IT*** with us, please let me know so that I can post ***IT*** on my web page for others to read.

In my quest to understand ***IT*** better as ***IT*** exists as omnipresent, the more I found out about the ways ***IT*** exists as hot and cold, as weight and nothingness, I have had to accept that the things ***IT*** does are incredible. I find the strange ways ***IT*** exists as what is known to science as pure energy and what ***IT*** does with ***ITS*** weight as ***IT*** reshapes. But still, there is no deeper satisfaction for me than the way ***IT*** exists as me and the way ***IT*** has allowed me to directly connect with ***IT*** through meditation, as ***IT*** exists inside of me as ***ITS*** nothingness, and how ***IT*** interacts with me and what ***IT*** exists as outside of my human body, in terms of objects, people, and events, as ***IT*** is reshaping ***ITSELF*** as this show that we are witnessing. I am deeply grateful that I have been permitted to observe while ***IT*** allows me to exist as just one moment of ***ITS*** existence, as a gift called "my life". However, I do not feel the same way when I merely use my intellect to think about what ***IT*** may be doing with ***ITS*** heated weight somewhere else within ***ITS*** freezing cold, nothingness shell body, for I know that my gift is not there as ***ITS*** weight. ***IT*** is here, wherever I am, that I can be directly with ***IT***, both as ***ITS*** heated weight, and ***ITS*** 95-99% nothingness which is the way I also exist.

The Big bang and distance

Let me also say that at the moment of the Big Bang, ***ITS*** two energies where already there. Let me explain what I mean by this: Science has postulated that what occurred at the moment of the Big Bang was that there existed one very dense hot spot, and ***IT*** exploded outwards. As this heat traveled outwards, ***IT*** cooled off to the freezing temperature that now exists as what is referred to as the cold, dark matter that exists as this empty Universe.

I should also clarify that I too used to believe in the theory that states that at the moment after the Big Bang, the temperature of the Universe cooled down, and the reason for my agreement was that I too was focusing on the heat that existed inside the Universe, for my mind also went straight to what I could see and understand, which was what ***IT*** existed as matter, while totally ignoring ***ITS*** clear, freezing cold nothingness. But now that we have more scientific information on how ***IT*** exists as the dual temperatures that exist as pure energy, we should revise our thinking about the Big Bang.

First of all, science has confirmed that this Universe is composed of a pure energy that cannot be created or destroyed and that exists as 2 extreme temperatures: first, as the heat that exists inside matter which makes up less than 4.5% of the Universe, which exists inside this second way that this pure energy exists in the form of a huge, freezing cold emptiness.

Now, keeping this in mind, let us return to the moment of the Big Bang when there existed one very dense, hot spot, that exploded outwards. There is no problem with this scenario. What I feel needs to be corrected is what happened as this heat went outwards and for the purpose of this discussion, let the reader remember that this pure energy we are talking about exists as 2 extreme temperatures as I stated above, and that these temperatures themselves as pure energy cannot be created or destroyed, for otherwise, atoms could not exist, and you and I would not be here.

My observations have shown me that this pure energy as heat can be fragmented but only within this pure energy's opposite temperature as coldness, such as the heat that exists inside each atom and as the heat that these atoms exist being inside of the freezing cold Universe.

It is important to keep in mind also that this pure energy as coldness has always been a constant, so that what happened was that the once singular heated weight that existed inside of this pure energy's other form of existing, that is, as a freezing cold

nothingness, was able to scatter this heat into smaller fragments, throwing them outwards, so to speak, into ***ITS*** clear freezing cold nothingness. In other words, what happened at the moment of this big explosion is that this pure energy sent ***ITS*** heated weight outwards (what we refer to as the Big Bang) in fragments into different parts of the same cold area that exists as this pure energy's huge, clear, freezing cold nothingness that has expansion or extension, but has no borders or boundaries. This is a quality that we are not yet familiar with. ***IT*** is difficult for our minds to conceive of something that exists as a freezing cold nothingness that does have expansion or extension but exists without having weight and has no beginning or end, as an opposite to this pure energy's other way of existing, as a heat that does have weight and exists as something tangible which we call matter.

I would also like to add that as I have tried to describe the way this pure energy exists in terms of dual temperatures and the way ***IT*** accomplishes things, I have also tried to convey that ***IT*** behaves in an abstract way, that is to say, without feelings or consciousness, and I admit, this is not the way most people conceive of God. But there is no doubt that scientists know that this pure energy does exist and that this Universe is totally composed of this pure energy (***IT***) that exists as a duality, in the sense of having two extreme temperatures that cannot be created or destroyed.

May my readers remember that you and I can only exist, see, think, and feel, because we are entirely made up of the dual way this pure energy exists. The fact that we have preferred to use the word God to designate this pure energy is not problematic for they are one and the same, because for God to exist inside of this Universe that we exist in, ***IT*** to has to be made of something, and for this God to be omnipresent, ***IT*** must exist inside this Universe which is composed of these two extreme temperatures, because there is nowhere else for ***IT*** to exist!

Finally, using our imagination we can see how the ending of our Universe will come about. When all of ***ITS*** heated weight is finely compressed into one last black hole that has consumed all other

black holes we will again be back to what we know as the beginning of the Big Bang, for ***IT*** is in the process of what black holes do in terms of "swallowing" and compressing matter that ***IT*** brings back all ***ITS*** fragmented, heated weight into one singular heated weight within ***ITS*** freezing cold shell body, so as to again throw out ***ITS*** heated weight reshaping ***IT*** into new possibilities inside of ***ITSELF*** as ***ITS*** freezing cold shell body, for ***IT*** cannot do anything outside of ***ITSELF***.

Gods body
Here is something that we very seldom think about: Does GOD have a body? And if so, how does this body exist?

Now I have also placed this subject in this section that has to do with matter, because ***IT*** is in this area, that we can scientifically see how GOD'S body does exist. To begin, let's first review what we do know exists scientifically about GOD, so that if GOD is everything that exists, then everything that is known to exist as this Universe, where something called pure energy exists, are one and the same. This reminds me of those Affidavits people sign when they are known by another name or alias. "I issue this Sworn Statement in good faith, so that all may know that GOD, also known as the Universe, are one and the same person." If GOD is every thing that exists, then this Universe also has to be GOD.

And people that see this Universe scientifically also know that this Universe exists because of something called pure energy. Now let me add that, since the moment our human minds began to exist, we have, for natural reasons, gotten used to understanding GOD more at the level of our existence, such as the idea of GOD being mostly here on this planet, even if we know that everything that exists outside of this planet exists as GOD, as being everywhere, or as being everything that exists. But ***IT*** was science that explored more of how this Universe exists in the form of pure energy.

So that for those people that know that God is everywhere, try this as a way to understand how God's body exists: As you look out at night into what exists as this Universe, remember that you are

looking out, for we as humans cannot stand outside of God and look into how God exists.

As you look into what exists out there as this Universe and keep in mind that everything that exists, exists as God, you will notice that this Universe is huge, to the human mind. So also, is God huge.

The first thing that you will notice is the many things that you can see, that do exist inside this Universe, but henceforth you will be more aware that what does exist out there as this Universe is really God's weight reshaping.

Then, as you keep looking into what is out there as this Universe, remember that you can only visually see how God exists as ***ITS*** weight, and you will also notice that this weight does exist, and that this weight that you are seeing exists in a place that does exist, known as a clear, transparent, empty, freezing cold nothingness that this Universe exists as.

The more you look into this Universe, bear in mind that the only reason it is possible to see the things that you do see is because you are looking at God's weight. Someday when we develop more powerful telescopes, we will be able to see even deeper as what exists as ***ITS*** weight as this empty Universe.

And when we develop bigger space ships, we will be able to travel more into how ***ITS*** weight exists, within ***ITS*** freezing cold, clear, empty nothingness as ***ITS*** body.

The reason I have made the preceding statements is because this is the only way we can understand that what we are seeing when we look into this Universe is really God's weight existing inside God's body, which exists as a freezing cold, clear, empty nothingness.

Now also remember that this freezing cold, clear emptiness does exist as being there, for we at least know that it is so, because it is this emptiness that permits us to see what God exists as from one weight (what we call matter), to another, so again we should

remember that from one weight to another something does exist in between, and if this freezing cold, clear, empty nothingness does exist as an energy, then this too is God, for nothing that exists as a freezing cold energy that exists as something can exist outside of God.

This being so, one must come to the conclusion that if God is omnipresent, what is out there as this Universe also has to be God. And if these things that exist out there are God, then this freezing cold, clear, empty nothingness that exists out there as pure energy is also God.

You may be wondering why I have placed this information here, within the section on matter. This is why: If God does exist, then God is something, and if God is something, then God exists as matter.

Now, if God exists as matter, then we should explore how God exists as matter, because we know that God is everything that exists. Furthermore, if we study God as matter, then we can use what we scientifically know about matter in terms of pure energy and more important yet, the "place" where this matter exists as pure energy.

Now scientifically, and obviously, since this Universe does exist, there are things that we already know about this Universe one of which is that this Universe exists as this pure energy, and scientists have discovered that this pure energy cannot be created or destroyed, and this is so because we cannot create or destroy this pure energy that God exists as.

So now, let's say you are a scientist, and you are looking at night into this place called the Universe, and using the information that we already have concerning pure energy, you would know that what you are visually seeing out there does exist, as the pure energy that matter exists as, and that this matter does have weight attached.

You would also know that this weight that exists as matter exists in a place that is referred to as being 95% freezing cold, empty space and that scientifically this 95% freezing cold emptiness does exist, and I will repeat this again, we scientifically know that this pure energy exists as 95% freezing cold, empty nothingness as the inside of this place called the Universe. There is no doubt that this freezing Universe as pure energy does exist, and in existing, it too has to be what is referred to as GOD, in the sense that God is everything that exists, for **God and pure energy exist in the same place.**

From the above the evident conclusion is that both GOD and pure energy, which I call ***IT,*** exist within one total freezing cold nothingness, and if we look at this pure energy or God as existing as what I refer to as ***ITS*** shell, or as the shell pure energy exists as, which behaves as an inner and outer shell simultaneously, and ***ITS*** weight, as God, or as pure energy, in the inside of this invisible, freezing cold, clear area that I call a shell, (which is not the best word, but will have to do until some one can see a more appropriate word or phrase to describe the way ***IT*** as God or pure energy exists as a body), where ***ITS*** heat is now found.

So now as a scientist, as you look into this Universe, you are aware that what you are seeing through your eyes is this pure energy, as the heat that exists as matter, that can only exist within this pure energy's empty, freezing cold nothingness. Therefore, if matter, or God, do exist, both have to exist within this freezing cold nothingness as a body that does exist in the form of this pure energy's body, because this is where this pure energy's heat is found existing within, so that both God and pure energy exist, and they both exist within this place, that is known as this freezing cold Universe, or omnipresence, and all these: God, pure energy, the freezing cold Universe, and omnipresence are one and the same.

I feel that it is proper, now that I am discussing the way God exists in this section on matter, that I should also address the subject of life as matter, as God, and as pure energy, let me also write you

about life as matter and relate this to ***ITS*** freezing cold nothingness.

Apologizing for my repetitiveness I would like to begin by referring to the way some people understand life, as only being given by God, or as something exclusively derived from God, about which you shall read more in later sections.

If we think of life itself as pure energy, it is very important that we should remember that pure energy **cannot be created or destroyed**. So it follows that life cannot be created or destroyed and also that life and God, as pure energy are really just one. This leads us to the next conclusion, that what we refer to as life only exists because there is something called pure energy or God, and these exist as one and they are not creating or destroying anything. They are only reshaping or transforming within themselves.

Putting all this together, we can say that ***IT***, as pure energy and God, exists as one being, that also exists as a place within which God, as pure energy can reshape ***ITSELF***, as what both are, just one, that is composed of two extreme forms of being simultaneously.

So, in order for life to exist, where we can see life, life has to have one very important factor, which is that all life forms have to have a fragment of God's weight, as the weight that exists in the form of pure energy, which cannot be created or destroyed. In view of this, we can then say that life exists where ***IT*** has taken ***ITS*** fragmented weight, that exists within ***ITS*** total freezing cold nothingness, and has given this fragmented weight mobility, and the appearance as existing as something that is alive. Now we can see life from a different point of view: If you remember that everything is ***IT***, as God or as pure energy, when you consider all the life forms that exist, you will notice that all of the things that are alive and moving, are alive as just one, and they are moving within ***IT***, as God or as pure energy. And the existence of all these things that are alive and moving is only possible because they exist within ***IT***, as God, as pure energy, as just one.

Let me discuss this further for the sake of clarity: If you take any life form, you will find that all of them contain ***ITS*** weight, as the different types of fragmented weight that now exist that came from when all ***ITS*** weight as pure energy was just one, at the moment of the Big Bang. So if we stop here for a moment and review the fact that ***IT***, pure energy or God, is not really creating or destroying, so that if ***IT***, God, pure energy, is just really one being, then you will see that, if life now exists, it is only because ***IT***, as God, as pure energy, exists as having this which is called life as ***ITSELF***, as God or as pure energy, because nothing is really being created or destroyed, as ***ITS*** weight or as ***ITS*** nothingness, or as ***ITS*** way of existing, as ***ITS*** way of being alive.

Now let us look at how ***IT*** as life, as pure energy is using ***IT***S once total weight, that is now fragmented, as what all life forms exist as. The process that took place after the Big Bang went something like this: ***IT*** took a tiny amount of ***ITS*** now fragmented weight, and became the electrons that all life forms need to have, and ***IT*** used this weight as electrons to encircle a other portion of ***ITS*** fragmented weight in the form of protons and neutrons, so as to exist as the matter that all life forms have, and it was ***IT*** that placed the mobility that ***IT*** as life has, so that when we see ourselves, as something that is alive, what we are really seeing is ***ITSELF.*** For example, if I am looking at a living person, what is really happening is that first of all, the person in question is there because ***IT*** took a tiny amount of ***ITS*** weight, and attached ***IT*** to ***ITS MAXX-SPEED*** so as to become electrons, and then, ***IT*** took other, larger quantifiable portions of ***ITS*** weight, that we call protons and neutrons, and surrounded these with the electrons ***IT*** shaped into so that what we call matter could exist, with a given distance with in ***ITS*** total nothingness, that we now see as the 95% empty space that exists inside the atom, because of the separation that ***ITS*** weight produced in the form of electrons, because the 95% emptiness inside that atom is the same empty nothingness that ***IT*** exists as ***ITS*** one total nothingness. Going back to the living person, what we see as the outer layers of the person's body are the electrons and the empty space that surround the protons and neutrons that all together make up the atoms that constitutes a

person's skin. These atoms, as you already know, exist because of the way ***IT*** is using ***ITS*** weight, (matter). Now let me return to ***ITS*** nothingness again, so that when you look at the person in our example you will be aware that what has happened is that ***IT*** took that very small amount of ***ITS*** weight in the form of electrons, and separated them by a distance from that other part of ***ITSELF*** that exists as ***ITS*** fragmented weight in the form of protons and neutrons that make up the atoms that form our bodies, and gave us what ***IT*** already exists as. I am not speaking about life now, but rather what ***IT*** exists as in terms of a divine consciousness, in the form of the nothingness that ***IT*** exists as that is always a constant.

To summarize then, the fact that we are alive began with the process through which ***IT*** gave ***ITS*** once total weight (the weight that existed at the moment of the Big Bang), fragmented this weight, in the form of electrons, protons, and neutrons, to put together atoms, so that ***IT*** could reshape into the matter that we exist as, and ***IT*** gave this weight certain functions, such as our brains, hearts, livers, and lungs, and certain functions that this weight could use for producing motion or mobility, so that ***IT*** could exist as ***ITSELF***, as a divine consciousness, which we see and call being alive. For all these reasons, when we see all of these human beings on our planet and everything else that exists, whether organic or inorganic, we should be aware that all of these things exist within ***ITS*** one total nothingness. Let us not be fooled by the impression that we are outside of something, for we are really still inside of ***ITS*** one total nothingness, and the reason is that ***ITS*** nothingness exists in everything that exists, as air, water, trees, houses, trains, airplanes, or everything and anything on and in this planet; all this is inside of ***ITS*** total nothingness, because anything that does exist, in order to exist, has to be made of ***ITS*** weight, which has to exist within ***ITS*** one total body that exists as what we see as this freezing cold Universe.

Finally, let me return to what I started out to say, which is that ***IT*** is very interesting, to see things as being out there, when in what we call reality everything out there is really *inside* of ***ITS*** one, total, divine, constant conscious, nothingness. All of this will make

more sense to you if you keep in mind that nothing is really ever created or destroyed, for everything that has ever happened or will happen will always be ***IT***S reshaping of ***ITSELF,*** as the way ***IT*** exists, as ***ITS*** two extremes. One thing that will help you visualize this is that it is the separation of ***ITS*** fragmented weight that gives us what we experience as physical distance within ***ITS*** one total nothingness, which likewise gives us the illusion of there being trillions or googols of things existing out there, when in reality, there is just one of ***IT***, where all ***ITS*** fragmented weight is moving within ***ITSELF.*** Furthermore it is ***ITS*** nothingness that is illusive, that produces the illusion that we see, because this nothingness, does exist, but cannot be touched or seen.

Nevertheless, it is not so important to me whether or not I understand life, as being grateful to ***IT*** for allowing me to be aware of ***ITS*** existence, for to me, there is no need to try and find the origins of life, as much as just being grateful to ***IT*** that ***IT*** exists as life and has allowed me to be here as one moment of ***ITS*** existence as life.

IT is

I have learned the following things related to that which we call the Creator, God, or Pure Energy, ***IT***: ***IT*** has expansion. ***IT*** is the outer borders of the Universe, which are light years away. ***IT*** has speed; 186,000 miles per second, where ***IT*** exists as a place called omnipresent. ***IT*** has weight; all the matter in the Universe is the weight of this God. ***IT*** has spin, necessary for the formation of atoms and for reshaping. In the spinning there is no past. ***IT*** is moving in one direction only. Spinning is what allows everything to exist at this immediate moment as a place.

Pure energy as positive and negative

For us to understand this pure energy further, we should remember that ***IT*** exists as two extreme forces, positive and negative. We have labeled pure energy as positive when it relates to the mass (weight) that ***IT*** exists as, and we can say that ***IT*** is negative as ***IT*** exists as the emptiness inside this Universe that does not have a beginning or ending.

So when we refer to pure energy as having no beginning or end, we are referring to the empty nothingness. Here, in ***ITS*** negative nothingness, ***IT*** is a constant, for this nothingness does not change. Rather, ***ITS*** weight changes, which is the same as reshaping, transforming, or transmuting ***ITS*** weight.

We can better understand this pure energy if we look at ***IT*** as two extreme forces. The reason for this is natural, for we have seen the infinite possibilities that ***IT*** can reshape into as ***ITS*** weight, as the matter that exists within this Universe within ***ITS*** constant nothingness. ***ITS*** weight is always moving and changing, for ***IT*** is looking to return to ***ITS*** original stage, which we can see in the way the force of gravity acts: pulling ***ITS*** weight as if to return to ***ITS*** original density the moment just before the Big Bang. So we now refer to the positive and negative of pure energy as being composed or labeled as follows: ***ITS*** weight being ***ITS*** positive and always reshaping, looking to return to ***ITS*** oneness as weight, and ***ITS*** nothingness as ***ITS*** negative, as ***ITS*** constant that exists as the huge, cold, clear nothingness of this Universe.

When we refer to pure energy as having no beginning or ending we should take this pure energy and divide ***IT*** into two. One is the 99.99% huge, freezing cold, clear nothingness as ***ITS*** negative that now exists as outer space, and ***ITS*** second part is the less than 1% weight, which we have labeled the positive and is constantly changing as ***IT*** reshapes into you and me and everything that now exists within ***ITS*** negative nothingness that we know as this Universe and can understand as a place.

You will understand this place a little better if you recall that you exist in a place that you call a town, and this town has to exist in a place called our planet, and that our planet exists in a place that we refer to as this Universe, so that every thing that exists as ***ITS*** weight will always exist in this place that exists as ***ITS*** huge, cold, clear nothingness, for ***ITS*** weight, as this Universe, cannot exist outside of ***ITSELF.***

One equals two

As I may have mentioned before, ***IT*** is one but behaves as two, as a duality of ***ITSELF***. Even in relationship to ***ITS*** nothingness, ***IT*** is constant as that part of ***ITSELF*** that is never changing. And as weight, ***IT*** is constantly changing as ***IT*** transfers energy.

⌘~~~~~~~~~~~~~~~~⌘⌘~~~~~~~~~~~~~~~~⌘

****** As ITS duality, one part of IT is constantly working to produce change(as weight or mass) and the other part of IT is at rest as a constant.*****

~~~~~~

IT as a constant

First, whatever ***IT*** is, ***IT*** is one, as in 100%.

Second, ***IT***, as one unit, is divided by ***ITS*** duality of two extreme opposites.

So let us review: The Universe, including us, consists of 5% matter and 95% empty space.

We know that matter has weight, and that all matter is ***IT*** as weight.

We know that weight has the ability to change and that ***ITS*** nature is to transmute.

I review so that we can better *see* ***IT***, and I stress the word *see* because we need matter (weight) in order to see things as manifestations of ***IT*** in ***ITS*** reshaping. ***IT*** has produced all that is necessary for us to develop methods to measure ***ITS*** speed and changing nature thus making us capable of comprehending ***IT***, even ***ITS*** more than 95% empty space, which is an integral and inevitable part of existence. If it were not for the empty space, ***IT*** would not have room in which to reshape. This 95% empty space is necessary beyond words.

Matter (weight) changes; emptiness is constant; this is ***ITS*** duality.

In the emptiness, 186,000 mps exists; we already accept this for we have the phrase, "time stops at the speed of light." Time and matter
~~~~~~

cannot exist at this speed.

Those of us who meditate already know that during meditation we connect with this part of ***IT*** that is constant and timeless. Furthermore, we humans are another of ***ITS*** extreme dualities. Our human body is in constant change and the opposite of this is life as ***IT,*** as that part of ***ITSELF*** which is constant. Also, this duality of ***IT*** can and has existed as pure energy or God.

Why water is not hot

Since atoms are made from fragments of ***ITS*** heated weight, why then, when I run the garden hose water, can I touch it to find it is not hot? Water is made of molecules that are composed of atoms, and atoms have heat, as the hydrogen and oxygen atoms that compose water.

If I could look inside the hydrogen atom, I would see that since it is the proton that has heat, it makes sense why the heat does not affect the temperature of my water. The proton is buffered by the distance of its outside environment-the 95% emptiness that exists in every atom. To this I can add that the electron does not make it easy for anything to get in or out of the atom unless the proper conditions exist that are related to physics and chemistry.

Here we see more clearly why at least 95% of each atom exists as nothing, for it is known that when the proton releases energy as heat, inside the atom, the electron absorbs the heat, which would likely energize the electron, but let me qualify my assumption: I am not a physicist or chemist or rocket scientist. I am just someone who loves to question my environment, for I have learned that questioning becomes the door that will open my mind to seeing the existing possibilities. I have found that the alternatives that arise from my questions empower my mind toward growth. I'm made to look for the thoughts, as information that I did not even know was inside of me just waiting to get out.

I can now see that writers, of which I am not one, or ever was, or ever will be, which I will explain later, will take a long time to

finish what they start. In my personal experience, when I begin to question something, my mind becomes active and when it comes down to putting it in written words, I have to re-learn what I mean in order to make my words as clear as possible so that eventually you, as a reader, can understand in sentences what started for me as paragraphs of questioning that end up as words that need to be reordered so that what started out as a thought to myself in the form of a question could be understood by you the reader as a clear idea.

It seems that a simple question to myself becomes bigger as I reach for the possible answers, bigger as in longer and longer, and not because I want to write more. No way! It is just that as I start to write my mind starts bringing more information that is related to the subject. It is then that I also realize that I, like you, will find that I have a lot of information stored inside of myself. When we start looking for information on a particular subject we seem to dig out all the other stuff that is there also, which we had forgotten was there, if we ever knew it existed in the first place. This is what happened with me: When I came in contact with ***ITS*** nothingness, the first question that came to me was, what is there that exists as information on ***ITS*** nothingness? And I was surprised to see that we have a lot of information on ***ITS*** nothingness.

And one more reason why I have said that we should look for what exists within us is because I am sure that there are readers out there that will have more information on ***ITS*** nothingness that can help contribute to understanding ***IT*** better.

ITS temperatures as how hot and cold IT is

Here is a question that some reader might be able to answer. The question is related not only to how ***IT*** exists concerning ***ITS*** size, but also how ***IT*** exists concerning ***ITS*** temperature, that is, ***ITS*** simultaneous extreme coldness and extreme hotness.

Let's begin by addressing ***ITS*** extreme coldness. Apart from ***ITS*** being as a nothingness, we could say that this extreme coldness occupies 99% of ***ITS*** area. For the purpose of this discussion, let us

say that in this 99% ***IT*** is hypothetically 100 degrees cold.

There is then a remaining 1% of ***ITS*** area which exhibits extreme hotness or concentrated heat. Again, for the purpose of this discussion, let us say that in this 1%, ***IT*** is 100 degrees hot.

Of course we know that ***IT*** can exist as heat that we can measure in the millions of degrees, such as our solar system's Sun, and ***IT*** can exist as absolute coldness, such as that encountered in the freezing cold hugeness of this Universe.

Having clarified that ***IT*** exists simultaneously as extreme coldness and extreme hotness, I propose that it is precisely these ratios of size and temperature that permit ***IT*** to exist as a oneness, as in ***ITS*** stage #3, a balanced state. In ***ITS*** huge, cold nothingness, where ***ITS*** temperatures reside, ***IT*** is able to take this 1% in which ***IT*** exists as extreme hotness and redistribute it into ***ITS*** remaining 99%, therefore existing at a neutral temperature, neither hot nor cold. This is precisely the balanced state that I refer to as stage #3, in which ***IT*** exists as one huge area of balanced temperature, inside that empty area that ***IT*** now fills completely which is the size of this Universe. I say size because ***ITS*** nothingness is also the area which ***IT*** occupies in terms of space. In these terms, size, heat, and distance are all interchangeable.

I know that it is hard for us to comprehend that ***ITS*** nothingness cannot be compressed, for how can nothing be compressed?

ITS cold hugeness is precisely where ***IT*** uses cold to redistribute the heat which exists within ***ITSELF,*** since cold has the property of pulling heat. And also because heat has the quality of movement, as in transferring and/or transmuting into infinitely small fragments, which is precisely the process that is now occurring within ***ITS*** cold hugeness.

So it may be true that this Universe, as weight, will become infinitely smaller, up to the point where this heat can be incorporated back into ***ITS*** cold, huge nothingness, thereby evenly

balancing ***ITS*** temperatures.

Remember that since ***IT*** is 99% freezing cold nothingness, if this weight that exists as heat were evenly distributed, there would be no problem of ***IT*** becoming unbalanced because ***ITS*** weight is only 1% and this weight can be fragmented into much smaller bits; bits even as small as the mass of a photon.

Now, if it were possible to calculate ***ITS*** heated weight, we would then know ***ITS*** weight as pure energy. This is the energy that exists now within this Universe, because even in ***ITS*** balanced state, ***ITS*** weight as pure energy must be conserved.

I personally would not want to be the one that has to find a scale big enough to calculate ***ITS*** weight, for ***IT*** is extremely huge, and ***IT*** is extremely heavy. Such a physical feat would be impossible, for the scale used would have to be ***ITSELF.***

How dense is heat?

How much can dense heat be reduced? Or let me present the question this way: Before the Big Bang ***IT*** existed as one very dense weight. Then ***IT*** threw this very dense weight outwards into very tiny, quantifiable portions of weight, as atoms. ***IT*** can make its weight even smaller, like the weight of the electron, or as the weight in a ray of light or a photon.

Now we know that ***ITS*** weight can be very heavy, as it was before the Big Bang. ***IT*** can reduce this heat, as weight, to minute fragments so that it can move about as the light rays of a star.
Let me give you another example of how ***IT*** can reduce its once very heated, dense weight. The light bulb is 4% light and 96% heat. So, it may be that the light that now exists in the whole Universe is near this ratio of 4% that ***IT*** exists as, as in ***ITS*** own image.

But back to other ways that ***IT*** has reduced ***ITS*** heated weight. We can see this with our bodies, as in the heat that we release from our skin, and from the ways we dispose of our body waste, just to

mention a few.

I have said the above so that we can understand that ***IT*** knows how to take its heated weight and reduce it into what now exists as ***ITS*** fragmented heat. ***IT*** can go from millions of degrees of heat, as in stars, to smaller fractions that are not as hot. To this I say "Thank you, ***IT*** for allowing me to question the various ways that your heat exists."

IT knows how to release its very dense, heated weight and fragment it into smaller fragments, as light rays, and eventually bring back all the fragments to a singular point, post Big Bang, and reshape ***ITSELF*** over and over again. This may not be the only time ***IT*** has done this. We should remember that ***IT*** has always existed, and ***IT*** knows what ***IT*** can do with ***ITSELF***, for ***IT*** is a divine form of consciousness.

Maybe you can visualize what I am saying, if I present it in a different way. Since ***IT*** took ***ITS*** very dense, heated weight, and threw it out as smaller fragments, ***IT*** can also bring it back.

Heat as the mover

It is ***ITS*** heat that we have been using as energy, to move things, because ***IT*** is heat that permits the things that we know as matter to move, and we measure ***ITS*** heat as caloric energy as the heat that all matter exists as, but ***ITS*** heat cannot be used without ***ITS*** nothingness, just like a battery. A battery has energy, but this energy cannot be released without the negative nothingness.

The energy that we are utilizing is from ***ITS*** heat, for ***ITS*** nothingness is still a constant. However, we must use ***ITS*** constant nothingness in order to be able to utilize the energy that ***ITS*** heat has.

So this does raise one question, at least in my mind: If it is ***ITS*** heat that has energy, what does ***ITS*** freezing cold, clear, nothingness exists as, being made of pure energy? Can anyone find an answer to this problem?

ITS heat within ITS coldness

Scientists have commented that ***IT*** existed at the moment of the Big Bang having a temperature that is estimated to have been in the trillions of degrees, as ***ITS*** heated weight, which may mean that every fragment of ***ITS*** weight will be lower in temperature, not because this fragment is truly, in itself, lower in temperature, but because this now smaller fragment is now surrounded by more of ***ITS*** coldness, that exists as one portion of ***ITS*** weight surrounded by ***ITS*** coldness, as the same coldness that exists outside.

Let me explain this, from a different angle. Let us say that when ***IT*** had all ***ITS*** weight in one place, as in the moment just before the Big Bang, ***ITS*** weight as heat could be kept at a higher temperature, in the trillions of degrees, because one singular point was easier to hold at this high temperature while being surrounded inside of ***ITSELF*** by ***ITS*** coldness.

Look at it this way: When ***IT*** reshaped ***ITS*** weight as the Sun, where ***ITS*** once singular weight that was at trillions of degrees now exists as fragmented weight, we can see that the temperature has dropped to only millions of degrees, and when this once trillions of degrees is lowered into fragments as small as protons, neutrons, and electrons it is easier to lower this temperature because it is surrounded by ***ITS*** freezing coldness, since it is now separated by ***ITS*** weight in the form of electrons, so that the heat that was once in the trillions of degrees, can exist as a lower temperature due to ***ITS*** coldness. This is another reason for which I have to say thank God, that ***IT*** did this, because at the present temperature we are able to exist, for it would be impossible for us to exist if things were otherwise.

IT will be easier to understand if we just take ***IT*** as ***IT*** exists. There is no doubt that ***IT*** exists in dual ways: As ***ITS*** freezing, cold shell body, where ***IT*** keeps ***ITS*** heated weight inside, and as ***ITS*** fragmented weight, which is the other way ***IT*** exists. Now, this being so, you, the reader should remember that both of these two temperatures, as pure energy cannot be created or destroyed.

What is happening is that ***ITS*** coldness is lowering this hot temperature by fragmenting it into smaller pieces, so that it will be easier for ***ITS*** coldness to lower this smaller fragment, surrounding it with more of ***ITS*** freezing coldness, for this temperature cannot be created or destroyed, since this is the law that governs this pure energy.

ITS freezing cold body as something
First, let me remind the reader, that we as humans can understand things because we have a mind that is capable of analysis. In the following paragraphs what I am trying to say is that for us to truly except something as a fact, it helps if this something does exist physically, palpably, in such a way as to be perceived by our senses. In other words, for our minds to accept something, it has to *be* something. Starting with this premise I will attempt to explain to you something about how ***ITS*** cold, transparent body exists. So throughout this exposition, please remember that for our minds to accept something, it should be made up of something.

We have information that this Universe does exist as something, because we can visually see that something does exist, which we call matter, which is really ***ITS*** fragmented weight. Thanks to the invention of instruments such as the telescope, we have been able to understand the way ***ITS*** weight exists as the matter that exists "out there", which is really inside the Universe. We have also found that 95% (I really should be saying 99.99%) is made up of what we have been calling Dark Matter, and in order to understand this huge nothingness that ***ITS*** body exists as, we have had to call it Dark Matter, in order to conceive of it as something we can hold on to as existing, because we know that this freezing cold, transparent nothingness has extension; that is, it is measurable in terms of distance.

Scientific evidence has also been uncovered which indicates that this freezing, cold, transparent nothingness also has sound, because our instruments, (which, by the way, are themselves made of ***ITS*** weight as matter, and therefore are 95% empty space if we consider the empty space inside their atoms), have detected radio

waves. Now, I do not know too much about ***ITS*** nothingness body, for I am only sharing with you what I have found, so that readers out there can use what little I have understood about ***IT*** as ***ITS*** weight that exists within ***ITS*** freezing cold, transparent nothingness body. I am sure there are minds out there that will uncover more aspects of ***ITS*** nothingness when they too focus on ***IT*** as this body that ***IT*** has, which exists as a freezing cold temperature that does have distance, as something for us to investigate, as something tangible. Hopefully we will be sending out space ships deeper into ***ITS*** inside that we from Earth are unable to perceive.

I feel that these radio waves our instruments have detected as existing inside ***ITS*** nothingness, are emitted by this speed that ***IT*** also exist as, that we call the speed of light. This speed is also an aspect of ***ITS*** existence, and as such this speed cannot be created or destroyed. so that before the Big Bang, when ***ITS*** weight was not yet attached to ***ITS MAXX-SPEED***, which was to become what we now call the speed of light, and as such, still has to exist. I feel these findings concerning ***ITS MAXX-SPEED*****,** which exceeds 186,000 mps that to me exists as ***ITS*** cold body, suggest that just as something that is traveling at high speed can make a noise as a sound, in this same manner these radio waves that we detect are coming from ***ITS MAXX-SPEED***. Remember that since this ***MAXX-SPEED*** does not have any of ***ITS*** weight attached, it is difficult to detect. Perhaps there is a reader who knows of an instrument that can detect speeds that are faster than 186,000 mps, while taking into consideration that the instruments that are being used to detect ***ITS*** transparent, speeded, freezing cold body will have to be made from both ***ITS*** dual temperatures: from ***ITS*** way of existing as the heat inside the atoms that make these instruments possible, and from the high speed that ***IT*** exists as the high-speeded electrons that are part of the atoms that make up the instruments' matter.

One last thing to remember concerning this is that as ***ITS*** freezing cold body does not consist of something, which is why this area cannot freeze up or cake up, so to speak, or become denser or

tangible, such as ice. However, using ice as an example, when ***IT*** reshapes into ice ***IT*** does so as ***ITS*** weight, as the weight that makes ice possible, and as always I have to say, no matter what we find concerning how ***IT*** exists as ***ITS*** freezing cold, transparent nothingness body, nothing will change the way ***IT*** is nor what ***IT*** will do with ***ITS*** weight that exists within ***ITS*** freezing cold, transparent body.

Let me also say that I have offered these thoughts on ***ITS*** different ways of existing, in hope that there are people out there that can add more to the information that I now have. I feel confident that this book can generate more interest on ***IT***, thus generating more information about ***IT.***

IT as hot and cold

In order to discuss this topic let's start with ***ITS*** heated weight as ***IT*** existed at the moment of the Big Bang. For simplicity's sake I am going to hypothetically posit a temperature to measure what existed as ***ITS*** heated weight at the moment just before the Big Bang, which I will refer to as +1000 degrees and conversely, I will use -100 degrees to address ***ITS*** freezing coldness. I am doing this so that we can try to keep track of the way ***ITS*** dual temperatures exist and behave. I say "behave" because ***IT*** is important to remember that everything that now exists as this Universe exists as pure energy, and this pure energy, when reduced to ***ITS*** basic form, exists as the 95-99% freezing cold nothingness, and within this freezing cold nothingness we find the other extreme, opposite temperature which is ***ITS*** heated weight. It is because of the way these two extreme temperatures interact or behave in relation to one another that everything that exists can in fact exist.

In my observation of this pure energy in this stage of ***ITS*** existence, I have seen that this Universe seems to be basically composed of a 95-99% freezing cold, huge, empty form of transparent nothingness. Since our eyes can see through this clear, freezing cold, huge, empty nothingness, we have tried to detect an outer border to it, but of course no such border exists, because how can nothing have a border? We have accepted that this area that

comprises 95-98% of the Universe does exist, but since it is composed of a clear, transparent, freezing cold way of existing, we cannot find a beginning or end to it. We do know that this area exhibits a freezing cold temperature, but we have not yet found a way to deal with ***IT*** because ***IT*** can't be touched or weighed. However, scientists do know that ***IT*** exists and accordingly, astronauts have to be protected by space suits and stay within the controlled environment of space stations and capsules.

We now have much more information on the way ***ITS*** less than 1% heat exists in the form of matter which comprises the 4.5% of the Universe (in terms of this pure energy), that exists at this opposite, extreme temperature, but still existing inside the other freezing cold, empty, clear 95-99% of the Universe. The reason for this is that our eyes can perceive ***IT*** and we can touch or feel ***IT*** as heat. This heat makes up the atoms that compose matter. In fact, we have had to use the way this pure energy exists as heat to build the instruments that we need just to confirm that this pure energy exists as a freezing cold temperature. But even the instruments themselves are made up of this pure energy in the dual manner of ***ITS*** existence: 1- as the heat from the atoms, and 2- as the empty nothingness that exists inside each atom.

⌘~~~~~~~~~~~~~~~~~~~~⌘⌘~~~~~~~~~~~~~~~~~~~~~~⌘

*** ***God and pure energy as omnipresent are the same.*** ***

❀~~~~~~~~~~~~~~~~~~~~❀~~~~~~~~~~~~~~~~~~~~~~❀

Perhaps I should explain this from a different point of view: Let's start with the fact that this pure energy exists as one, by which I mean that there are not two pure energies, but this one, total, pure energy exists at two independent, extreme temperatures. This is the same way this pure energy existed at the moment of the Big Bang. However, since the moment of the Big Bang, which is just this same, omnipresent moment, this pure energy has used ***ITS*** heated weight, as fragments within this pure energy's freezing cold temperature, to become atoms, so as to be able to become matter, in order to finally become everything that now exists as fragments of this pure energy's once singular weight. Now, even if we find a way to measure this pure energy's freezing cold temperature, we can only do this using both of these dualities that already exist as

part of our measuring instruments.

However, whether or not we can ever find out more about the way this pure energy exists as just a freezing cold temperature, I have to be grateful to ***IT*** for permitting me to at least know that this pure energy does exist, and that ***IT*** reshaped into me as I presently exist, so that I could somehow enjoy this pure energy's existence through seeing, hearing, touching, feeling, thinking, just to mention a few of the endless gifts in which this pure energy exists. So I would like to say to this pure energy: Thank you for giving all of these qualities to the human race. Nevertheless, even though I risk sounding selfish, the most important thing for me is that I have been allowed to deal with this pure energy directly. I learned that my relationship should only be with this pure energy, that we also refer to as God, and that I refer to as ***IT***, without any mediators or middlemen. The good news is that this kind of direct relationship with ***IT*** is not exclusive. ***IT*** is open to everybody and anybody, ready to receive and help anyone who asks.

⌘~~~~~~⌘

****** Since IT is in all places at the same moment as omnipresent, then IT is all ITS freezing cold nothingness and IT is also inside of ITS heated weight. ******

❁~~~~~~~~~~~~~~~~~~~~~~~~❁

So let us start with one of ***ITS*** dualities: ***ITS*** heated temperature that has weight, as one, total, singular weight, the way ***IT*** existed at the moment of the Big Bang, and since it is considered in the scientific community that this singular weight was billions of degrees hot, for the sake of simplicity let's make believe ***IT*** was +1,000 degrees. On the other hand, also for simplicity, I will hypothetically say that ***ITS*** opposite way of existing, that is, ***ITS*** freezing cold body made up of a nothingness which constitutes 90-95% of ***ITS*** totality, is -100 degrees cold.

So let's approach this scenario by saying that hypothetically ***ITS*** heated weight (which is less than 1% of ***ITS*** totality) can exist up to a maximum temperature of +1000 degrees in terms of heat and this heat, as pure energy, cannot be created or destroyed.

Now, using one's imagination, if we could have been near ***ITS*** one, singular heated weight at the moment of the Big Bang, and knowing that ***ITS*** heated weight existed then at +1000 degrees, which is ***ITS*** hottest, it is obvious that this +1000 degree heated weight (which is less than 1% of ***ITS*** totality) could and can only exist within ***ITS*** 99% freezing cold, clear, transparent body, which is precisely what we see now as this empty Universe.

Let me take a moment to clarify something: We have become used to saying that 95% of the Universe is composed of a cold empty place, and inside this place the remaining 4.5% is composed of matter. There is no problem with seeing the Universe this way, but in order to understand this pure energy better, we first have to accept that this pure energy is really 100 % ***ITSELF***, as in there being one God as 100%. Now, this 100% God, as pure energy does have certain obvious qualities, one of which is that ***IT*** exists at two extreme temperatures, and the other, is that ***IT*** also exists as what we have found to be 4.5% matter. This matter can only exist because it is made of atoms and these atoms have heat inside of them, and all of this matter exists inside a huge empty area that is composed of a clear, transparent nothingness, that exists at an extremely low temperature. We have chosen to call this "structure", so to speak, "the empty Universe". Nevertheless, one must remember that in using these ratios of 4.5% to 99% and +1000 degrees of heated weight (matter) residing within the -100 degrees of cold nothingness, we are speaking of only ONE pure energy that exists at two extreme temperatures and which at the same time comprises all the intermediate temperatures.

Maybe it will be easier to understand if I take you, the reader, as an example: You exist as one, as a 100% totality that is yourself, but your existence is only possible because you have the same qualities as this pure energy: You are made of matter and this matter exists as the heat present in each one of the atoms of your body, but you also exist as the 95% nothingness that can be found inside each atom. I say this because this pure energy exists as one, as the totality that is ***ITSELF***, but ***IT*** too exists as these two extremes: as heat and as cold, empty nothingness. Furthermore, our existence is

only possible because ***IT*** exists as this duality. Our existence was made possible when ***IT*** took a part of ***ITSELF*** to become electrons, to hold in ***ITS*** heated weight as protons and neutrons in order to form atoms. So you see, if it were not for the way ***IT*** reshaped ***ITS*** dual temperatures, within ***ITSELF***, we would not exist as the gift we are in terms of these two extreme temperatures.

Thinking about this, ***IT*** seems to me that this is what may be occurring: Wherever these two extreme temperatures encounter each other there should be a difference between ***ITS*** -100 degrees, and ***ITS*** +1000 degrees only around the circumference of ***ITS*** weight.

You might better understand the above by using the common hydrogen atom as a model. You will see the same situation where, let's say, again hypothetically, the proton, as a fraction of ***ITS*** once total heated weight, is +1000 degrees hot (in the sense that ***IT*** has conserved the same temperature ***IT*** had before the Big Bang, but now, being just a tiny fragment of ***ITS*** original weight, ***IT*** will exhibit a milder temperature) but ***IT*** will still have to exist within ***ITS*** -100 degree cold body, which is our empty Universe.

Now the big difference that exists between the situation before the Big Bang, when ***IT*** had all ***ITS*** heated weight in one place, is the difference in temperature that existed at the exact place where ***ITS*** +1000 degrees came in contact with ***ITS*** -100 degrees. After the Big Bang, ***IT*** fragmented ***ITS*** one, entire heated weight that existed at +1000 degrees into smaller portions, which became all the fragments that exist inside the nuclei of all the atoms that exist (protons and neutrons), but these portions of ***ITS*** +1000 degree temperature are still existing within ***ITS*** -100 degree cold body. The only difference is that ***IT*** took another fragment of ***ITS*** heated weight and endowed ***IT*** with ***ITS*** high speeded nothingness to become electrons, and set these around the protons and neutrons, but separated from them by what now exists as the 95% empty space that is found in every atom, which is made from the same 95-99% cold, empty nothingness that ***IT*** exists as in deep space.

This fragmented +1000 degree heat that exists as the protons and neutrons inside each atom's nucleus is separated or is held at a distance because of the way ***IT*** placed ***ITSELF*** in the form of electrons which reside within ***ITS*** same -100° nothingness, and is now locked inside the nucleus, so to speak, so that the area occupied by the +1000° heat in the form of protons and neutrons is less than when ***IT*** had all ***ITS*** +1000 degree heat in one place occupying a bigger area as things were before the Big Bang occurred.

It is at the edge or circumference that there will be a temperature less than 1000° but hotter than -100°, for all the temperatures that can exist must be produced by ***ITS*** -100° nothingness meeting with ***ITS*** +1000° heat in the form of the pure energy that ***IT*** exists as, which as science has discovered, cannot be created or destroyed. If this were not so ***IT*** too would be destroyed in this destruction of ***ITS*** heat as weight or ***ITS*** nothingness. having a temperature of -100°. And then what would happen to ***IT*** in this process of destruction, not to mention what would our end be?

Furthermore, if something such as this happened, ***IT*** would never again be able to reshape ***ITSELF*** into the heat that a black hole exists as, in order to become again one, singular, dense weight, so as to return once more the elements that are needed to exist as a new universe, in terms of ***ITS*** weight only, that would have to exist within ***ITS*** constant, clear, never-changing -100° body.

Now, I have asked myself why temperatures such as +1000 degrees and -100 degrees do not exist on this planet and as far as I can see, the reason for this is that ***IT*** separated us in terms of distance by means of an atmosphere from what we call outer space where this -100° area exists. Or to state it a little differently: ***IT*** placed ***ITSELF*** in the form of trillions of atoms of ***ITS*** heated weight that exist as our atmosphere, so that ***IT*** would be a buffer between us and ***ITS*** -100° region in what we call outer space.

In addition, ***IT*** keeps us warm with the heat ***IT*** sends us as from the Sun in those areas of our planet that are able to receive ample

portions of ***ITS*** heated weight as energy, such as the tropical countries. We can feel this heat that ***IT*** sends when ***IT*** comes in contact with nerve receptors for temperature located in our skin. Likewise, we are able to get a taste of ***ITS*** coldness the closer we get to the North and South poles.

For the next step in our visualization, let's conjure a hydrogen atom in our minds. The structure of a single hydrogen atom consists of a fragment of ***ITS*** heated weight in the form of a proton and a single electron. These two fragments of ***ITS*** heated weight are separated by the same empty nothingness that exists in every atom. That understood; let us return to the issue of temperatures. In this single hydrogen atom, the proton, which is a fragment of ***ITS*** original +1000 degree heat, is surrounded by a portion of ***ITS*** -100 degree empty nothingness that separates ***IT*** from the atom's single electron. This separation would seem to allow for intermediate temperatures inside the atom which would be less than +1000 degrees but more than -100 degrees. This means that the part of the empty nothingness that is closer to the proton will be warmer or hotter than the part of the empty nothingness that is father away from the proton. Conversely, the part of the empty nothingness closer to the electron will be cooler or colder than the part closer to the proton. Just keep in mind that everything: the proton, the nothingness, and the electron, whatever temperatures they may exhibit are all ***IT***.

The discussion of the hydrogen atom above can be used as an analogy to understand ***ITS*** dual temperatures in the Universe. Before the Big Bang all the matter in the Universe was concentrated into a single point. When the Big Bang occurred, ***IT*** threw this matter outwards and in doing so ***IT*** became fragmented, which is why now we have discrete chunks of matter moving around within ***ITS*** empty nothingness. For the sake of simplicity, I have assigned a temperature of +1000 degrees to all ***ITS*** matter and a temperature of -100 degrees to ***ITS*** empty nothingness. To apply the analogy then, in the same way intermediate temperatures can exist inside a hydrogen atom, so there are intermediate temperatures between the chunks of matter and the surrounding

areas all the way to the freezing cold of ***ITS*** empty nothingness. In short, both of ***ITS*** extreme temperatures can and do exist within ***IT*** simultaneously, while at the same time the analogy allows for the entire range of intermediate temperatures to exist within ***ITSELF***, also.

Now I would like to attempt to explain why our atmosphere is warmer than outer space. To begin, please remember that our atmosphere is made up of several different elements in gaseous form and there are trillions and trillions of atoms as these gases surrounding our planet. Even though the atoms in gases are more separate from each other than in liquids or solids, they still are, so to speak, back to back or wall to wall, connected to each other closely enough to leave no space for ***ITS*** freezing cold emptiness as an area by itself as ***IT*** exists in outer space.

Before I go any further, let me tell you, the reader, something: I am no expert in this field of physics or cosmology. All that I am attempting to do is understand ***IT*** better, regarding the way ***IT*** exists. In this piece I am focusing specifically on ***ITS*** dual temperatures. I never had any contact with the information I am setting forth here before I started writing this piece, and I too have had to question certain information proposed by scientists. Let me explain why: I think ***IT*** is important to find out what is the temperature of ***ITS*** freezing cold nothingness. However, we may not have a way to really know what the temperature of ***ITS*** heated weight was at the moment of the Big Bang, even though scientists say ***IT*** would have been in the trillions of degrees. Could this just be pure conjecture?

On the other hand, maybe I am wrong about ***ITS*** heated temperature, and the only people that can give us a more exact measurement of ***ITS*** heat are scientists who have the necessary instruments or methods to in fact measure what is the true internal temperature of protons. I use the plural word, protons, because I am thinking of comparing temperature measurements of protons inside the atoms of the various elements. I guess I am just curious. Would the temperature of a hydrogen proton be the same as a

proton in the nucleus of an atom of gold?

Nevertheless, knowing that atoms have heat inside, in the form of protons, this heat may not exist uniformly throughout the inside of the atom. Let me explain myself: I propose that the heat from the circumference of the proton (obviously assuming that protons are "round") as +1000 should be hotter near the circumference, and as one moves out towards the electron, the temperature should be cooler, in keeping with the temperature of ***ITS*** -100 degree nothingness.

However, we should also remember that ***ITS*** +1000 degree temperature only makes up less than 1% of the atom, so that the +1000 degrees will be released only near the circumference of the proton and this heat will be absorbed by that part of the atom nearest to it which constitutes more than 99.99% of the entire atom. And if this is so, ***ITS*** +1000 degree temperature has remained constant and it is only what stems from ***ITS***-100 degree zone that has become warmer because it is nearer to the proton that is making contact with ***ITS***-100 degree area, that is, with the empty nothingness that exists inside each atom.

On the other hand, this too maybe wrong. Let me explain why: Scientists state that 90% of the Universe is made up of hydrogen atoms and within a hydrogen atom there should be a difference between the proton's heat and the electron's heat. Of course I have no idea what the temperature of an electron is, but since for the sake of this discussion I must use a number, let's say it is -750 degrees, since that number is less than -100 degrees, which would be the temperature of the empty nothingness that separates the proton form the electron. Now, if 90% of the Universe is made up of hydrogen atoms (which we cannot see because they are transparent) this would mean that ***ITS*** actual -100 degree, freezing cold shell contains 90% hydrogen atoms. The next logical step is to think that ***ITS*** freezing cold shell will be affected by the heat of these hydrogen atoms.

Looking at ***IT*** from a different angle: If we could collect all the heat that exists inside ***ITS*** -100 degree shell body, including this 90% made up of hydrogen atoms, plus all the other atoms that make up all the other matter in the Universe, and bring it back to a single point having a hypothetical temperature of +1000 degrees (the way things were before the Big Bang), we will have, in theory or imagination, "removed" the heat that was fragmented and scattered throughout ***ITS***-100 degree area. It stands to reason that having collected all these heat fragments, the temperature of ***ITS*** shell body would descend to something much lower than -100 degrees and the only area that would be radiating +1000 degrees would be this one place where all ***ITS*** heat is concentrated and the area surrounding ***ITS*** immediate circumference.

In order to question the scientific statement that ***ITS*** density at the singularity that existed before the Big Bang was tons per square inch I performed an un-scientific experiment that I described in the section called How dense are you? My conclusion was nowhere near what I found by mentally removing ***ITS*** nothingness. Instead, the result I came to was that ***ITS*** heated dense weight would have been more like 4-6 lbs per square inch.

On the other hand, all these calculations really make no difference to ***IT***, for ***IT*** will continue to exist as ***IT*** exists. The only difference would be our understanding of ***IT*** in terms of how humans exist for the time being as ***ITS*** heated weight. I say "for the time being" because a moment will come when humanity will cease to exist. There is no doubt that our Sun will die, which will entail the destruction of our planet. Or if we have left this planet, at some moment disaster may occur. One of the possibilities is that of being consumed by a clear hot spot (what scientists call a black hole) whether we are on a spacecraft or on a planet located in the nearby area.

Another question that comes to mind is: What if ***IT*** does exist at a temperature between ***ITS*** -100 degrees and ***ITS*** +1000 degrees? This could be construed as a neutral balanced temperature. But about this, we still have no knowledge. We could safely say that

we do have knowledge about ***ITS*** -100 degree temperature, out there in the empty nothingness of space, but we really do not know what the maximum temperature of ***ITS*** heated weight could be. As an example, if the cold temperature of ***ITS*** shell body is-100, then for ***IT*** to exist as at balanced temperature; ***ITS*** heated weight would have to be +100.

Now since we do have some information on ***ITS*** dual temperatures and considering that ***ITS*** coldness makes up 99.99% of ***IT*** as empty space while ***ITS*** heat makes up less than 1%, maybe a physicist or a mathematician can work with these numbers and find out what could exist exhibiting these two temperature extremes simultaneously, while at the same time remembering that these two temperatures as the pure energy that they exist as, cannot be created or destroyed.

Now, when a proton decays into smaller fragments of ***ITS*** + 1000 degree heat, these smaller fragments are surrounded by more of ***ITS***-100 degree coldness, the same way all the subatomic particles that exist inside the atom also exist at +1000 degrees, having less of ***ITS*** weight, but are surrounded by more of ***ITS***-100 degree coldness. Furthermore, protons can decay by losing energy which results in their having less of ***ITS*** heated weight which can move about within the atom, most probably accelerating up to ***ITS MAXX SPEED*** and being transported to some other location within ***ITS*** nothingness as ***ITS*** weight, for ***ITS*** weight and ***ITS*** nothingness, as pure energy cannot be destroyed; only reshaped, or transmuted by adding one fragment of ***ITS*** weight to an other fragment of ***ITS*** weight, by means of ***ITS*** speeded nothingness (***MAXX- SPEED***).

For all these reasons, I have come to the conclusion that any temperatures that exist come from ***ITS*** two primary temperatures, one as ***ITS*** -100 degree area, which is what allows ***ITS*** +1000 degree area to warm the -100 degree area closest to the circumference of ***ITS*** +1000 degree area, for these in between temperatures are temporary, because as soon as ***IT*** becomes a clear hot spot (a black hole - which is the way ***IT*** uses to bring back ***ITS***

fragmented +1000 degree heated weight in preparation to returning all ***ITS*** fragmented +1000 degree weight to a singularity such as existed before the Big Bang) the once warmer areas that existed surrounding them will return to a temperature of -100 degrees. This will happen on an atomic scale as well as on a cosmic scale.

Most of the Sun's heat, (in terms of its 10^{57} or so atoms are hydrogen atoms), is closer to ***ITS*** +1000 degrees. Fortunately, as the Sun's radiant energy travels through space towards the Earth ***IT*** loses heat, or else we would be in deep trouble. Maybe at some moment in the existence of humanity we may find a way to the Sun without suffering the effects of ***ITS*** searing heat. This way we could measure just how far out ***ITS*** heat warms up the area around ***ITS*** circumference. However, there will come a moment when our Sun explodes, becoming a red giant in preparation for collapsing into a black hole. When this collapse takes place, ***ITS*** circumference will be smaller and denser, and will have a greater gravitational pull on the surrounding area. It would be logical to conclude that the surrounding area will be colder than it is today.

I have thought of another way of looking at the way ***ITS*** +1000 degree heat behaves, both inside the Sun and throughout ***ITS*** entire freezing cold body: Every time ***ITS*** +1000 degree heat comes in contact with ***ITS*** -100 degree coldness what happens is that the +1000 degree heat warms up the surrounding area that has a temperature of -100 degrees. In the case of our Sun, when ***ITS*** +1000 degree fragments leave the surface of the Sun and begins traveling at 186,000 miles per second through ***ITS*** -100 degree area, this temperature drops, so that by the time they reach the Earth 8 minutes later, we receive this radiant energy at different temperatures depending on which area of the planet is targeted.

Perhaps it will be easier to understand this situation better this way: Every fragment that came from ***ITS*** original +1000 degree heated weight, still retains this temperature but in a smaller portion as heated weight, so that, for instance, the heated weight that exists as the hydrogen proton in each hydrogen atom is just one portion

of ***ITS*** original +1000 degree heated weight. Every atom of helium contains two portions of ***ITS*** original +1000 degree heated weight as 2 protons, plus another portion as a neutron, and all these 3 particles are still surrounded by ***ITS*** -100 degree coldness, which is found in the emptiness present inside every atom.

⌘~~~~~~~~~~~~~~~~~~~~~~⌘~~~~~~~~~~~~~~~~~~~~~~⌘

** ***Calling all thermonuclear minds! Can you help out???*** **

~~~~~~~~~~~~~~~~~~~~~~~~~~~~

I have some additional thoughts to share relating to ***ITS*** -100 degree coldness and ***ITS*** +1000 heat, but before we delve any deeper into this, we need to remember that this pure energy has 2 qualities: one, ***ITS*** heated weight, that can exist in a single place, and two, ***ITS*** freezing cold nothingness. However, we also need to keep in mind that this nothingness that still exists today as the empty freezing cold nothingness of the Universe is also a part of this pure energy, and ***IT*** has the property of extension or expansion; that is, ***IT*** occupies physical space, and furthermore, since ***IT*** is made of this same pure energy, ***IT*** cannot be created or destroyed. In fact, this freezing cold nothingness must exist in order for this pure energy to keep ***ITS*** heated weight inside of ***ITSELF***.

So now, returning to the moment of the Big Bang, let us bear in mind that this heated dense weight at existed so tightly compressed, that scientists affirm that it had a density of tons per square inch, at one singular spot.

From here, this heated weight was thrown outwards, and it is here that we have to use some imagination, because for this heated weight to move outwards it would mean that the space where ***IT*** was going to move into outwards already had to exist. If not, there would have been no place for ***IT*** to move. Therefore, the part of this pure energy that existed at the moment of the Big Bang for ***ITS*** heated weight to move into outwards has to be the same freezing cold nothingness that still exists today.

Looking at this another way, this heat that was moving outward had to have this freezing cold, empty space already available. Let's
~~~~~~~~~~~~~~~~~~~~~~~~~~~~

imagine that there is an area of 100 square miles where this pure energy exists as a duality, that is, as heat within a freezing cold nothingness. Following the Law of Conservation of energy which states that energy cannot be created or destroyed, then, this very area where this pure energy exists cannot be destroyed either, for there is no other "place".

So that, if we now have, let's say hypothetically, 100 miles of this -100 degree area existing as a freezing coldness in terms of area, then this 100 mile area would also have had to exist when ***IT*** had all ***ITS*** weight in one place, as in the moment of the Big Bang. This brings us back to the question of where this area came from if it was not already there within the original -100 degree area.

I say this because the heat that now exists inside this Universe which is at -100 degrees is now shielded as direct heat by the electrons, except for the heat that is emitted by the stars, but this heat is just being transferred somewhere else as the heat that light is carrying somewhere else within this pure energy's freezing cold nothingness.

I think we can better understand the moment of the Big Bang in the following way: Just before the Big Bang was the moment when ***IT*** had all ***ITS*** weight in one place within the same -100 degree area that exists as ***ITS*** shell body, which exists as pure energy, and as such, this shell cannot be created or destroyed, and when ***IT*** ejected ***ITS*** one singular weight into fragments, ***IT*** did not destroy this energy as heat, for this less than 1% as heat still exists as the 4.5% of all matter that still exists within ***ITS*** shell body, as this freezing cold Universe, as ***ITS*** dual way of existing. In addition, I should mention that if ***ITS*** heated weight is less than 1%, and this heated weight now exists as 4.5% which is all the matter that exists in the Universe, this would mean that ***IT*** used approximately 3.5% of ***ITS*** nothingness to form this 4.5% that exists as the atoms that make up all the matter that now exists within this pure energy's shell body.

Now I would like to try to explain that this area that exists as this

Universe, as space, or expansion or extension, that exists as pure energy, or as God, is the same area that this omnipresence has to occupy. Going back to the hypothetical 100 mile area, if this distance were reduced to let's say 90 miles of this energy that exists as a freezing coldness, where would ***IT***, as pure energy, move these 10 miles of freezing coldness? ***IT*** is obvious that ***IT*** cannot take ***ITSELF*** as pure energy somewhere else that does not already exist as ***ITSELF***, or as omnipresence.

Or we could try it in reverse: Let's say that at the moment of the Big Bang, when this +1000 heat occupied a smaller area in terms of expansion, and then, as this heat went outwards, where did this other area of "space" come from? Imagine that you were looking on at the moment of the Big Bang, and you saw the heated, dense weight about to explode outwards, then where is this "outwards", if it is not already there as this pure energy's dual way of existing as the opposite of ***ITS*** heated weight, which is none other than the freezing cold extension that already existed as ***ITSELF***? Otherwise, this pure energy would have had to create this freezing cold expansion. And where would this pure energy acquire this freezing cold extension that was not part of ***ITSELF*** as ***ITS*** own way of existing?

Let me mention that it is natural for us to just see the things that ***ITS*** heated weight exists as, as our primary way of understanding things that exist as matter, because our parents do not explain to us that there is something that exists as pure energy in the form of a freezing cold nothingness, that makes this Universe possible in terms of extension, within which anything and everything that exists as matter must exist.

Neither are we told that the only thing that we will find that exists as pure energy as omnipresent is the expansion or extension that this freezing cold pure energy exists as, because from the moment we are born and we open our eyes, the first thing we see is the way ***ITS*** heated weight exists, and when we first touch something, it will be ***ITS*** heated weight. This is due to the way ***IT*** has reshaped ***ITS*** fragmented, heated weight into the things we are seeing and

touching from the moment we are born. In addition, it is due to our programming, first for survival, that we are forced to focus on the way ***IT*** exists as ***ITS*** weight as the material things that we need, such as food, clothing, and shelter, so that we can continue to our second programming, which is reproduction. After reproduction our attention is focused on making sure that our children make it to where they are safe enough so that they can survive independently. When this period comes it becomes easier for some of us to be able to give more of our undivided attention to ***IT***, unless we decide to become a holy person such as a monk or a priest, who can bypass his/her second programming.

Each day our focusing on ***ITS*** weight starts from the moment we wake up. As soon as the proverbial alarm clock goes off we get attached to the things that ***IT*** exists as, that is, the objects that exist outside of us and other people. In fact, even when we are asleep, our mind continues its attachment to the way ***IT*** exists as the objects that we dream about, for our mind does not readily want to let go of the way the mind exists as our master, that has been in command from the moment we are born.

In spite of the fact that the mind is forced into having to rest during sleep, even during sleep the mind continues its function, which is to never shut down. So, the mind entertains itself with itself, in the form of thinking, which we call dreaming, which the mind produces in order to continue its existence, staying alert in order to fulfill its programming. It is interesting to notice that at least in this stage of the mind's existence, it does not produce any effects on the way ***ITS*** heated weight exists outside of our brains and bodies, until we naturally wake up again. I cannot say that the mind wakes up because the mind really never sleeps nor shuts down. Rather it is just that the mind is once again given the chance to participate in how ***ITS*** weight exists outside of our bodies. The mind just returns to seeing how ***ITS*** heated weight exists and to how the mind thinks ***IT*** is and also to who we think we are. Let me give you my personal experience, for I too had attachments to the things that exist outside of me as matter, but now I no longer have this type of attachment, for I, or my mind, has accepted that I, as my whole

body and everything that is “out there” is really just ***ITS*** heated weight reshaping. This means that my entire self and very existence really depends on ***IT*** as every atom that makes my body function, and as everything that exists outside of my body, for I have accepted that ***IT*** is 100 % everything that exists as ***ITS*** heated weight, which likewise exists inside of ***ITS*** divine, clear, transparent, freezing cold body.

⌘~~~~~~~~~~~~~~~~~~~~~~⌘⌘~~~~~~~~~~~~~~~~~~~~~~⌘

******Since IT is in all places at the same moment, then IT also has to be inside of ITS heated weight.******

~~~~~~

Now I feel very content and fulfilled, for I know I am in the safest hands that can ever exist, for I know that ***IT*** is the giver and taker. So again I say thank you to IT, for allowing me to connect to ***ITSELF*** in a direct way through meditation. I should also explain that this connection that happens during meditation can only exist by me wanting to be with IT, for this is a relationship that cannot be forced on a person; it has to come from my desire to be with ***IT***. Anything that happens from there on is only between me and ***IT*** and this is a gift that ***IT*** is allowing me to enjoy as the best thing that I now have, as my moment of existence.

Now let me mention that, as we deal with this pure energy that exists in terms of ***ITS*** temperatures, it may seem a little unimportant. Why bother with God as pure energy in terms of heat and cold? The way I see it, by understanding this pure energy that basically exists as just these two extreme temperatures we might be able to see and understand more of that which made us and everything else that exists, possible. Stop and think about this: You, I, and everything else that exists inside this Universe, exists as these two basic temperatures, at this stage of ***ITS*** existence and also at the stage of the Big Bang, for in both stages of ***ITS*** existence as this pure energy, both these temperatures have to exist, for both these temperatures are made of this pure energy that cannot be created or destroyed.

Perhaps the matter will be clearer if you look at ***IT*** this way: The dense heated weight that existed at the moment of the Big Bang is
~~~~~~

the same heated weight that now exists as you and I, and everything that exists as matter in the Universe, that still exists inside of this pure energy's freezing cold nothingness, for these are the two basic energies that ***IT*** as pure energy is made up of: 1) freezing coldness and 2) the heated weight that exists inside of this freezing coldness that has extension and can be conceived of to have a measurable width, within which this pure energy can use ***ITS*** heated weight to reshape into material forms that move about inside ***ITSELF.***

Another way to understand ***ITS*** dual temperatures would be in terms of ***ITS*** heat, but always keeping in mind that as this pure energy, be ***IT*** cold or hot, these dual temperatures cannot be created or destroyed. If we take the heat that this dual energy exists as, using a hypothetical temperature of +1000 degrees, and we went back to the moment of the Big Bang when this temperature existed as one, singular, dense, heated weight in one place, what happened was that ***IT***, as pure energy, threw out this heat in fragments into ***ITS*** other area, which had a temperature of -100 degrees. This is the way ***IT*** exists now, as the matter that exists inside this Universe. When this heat is brought back into being one, singular heat again, through the process of clear hot spots (black holes), this heat that now exists in these clear hot spots will have to be the same +1000 degrees that existed when this pure energy had all ***ITS*** heat in one place, for this heat cannot be created or destroyed. For this reason, when this heat existed in one place as +1000 degrees, ***IT*** was these same +1000 degrees heat that were fragmented, and these fragments, when added up all together, will still have to exhibit a temperature of +1000 degrees. Look at ***IT*** this way: As these +1000 degree fragments went outwards, they were individually separated by ***ITS*** reshaping into electrons. Now these individual +1000 degree portions are only warming up their surrounding circumference areas, and since ***ITS*** coldness cannot be destroyed either, this means that ***ITS*** coldness can absorb ***ITS*** heat to become temporarily warmer. I say temporarily because as soon as ***ITS*** heat returns to one singular place again, the areas that were warmer than -100 degrees will become cooler again.

So, as we have seen, scientists have completely accepted that this heated weight exists as the pure energy that is now present inside our Universe. Scientists have also accepted that an opposite to this heated weight exists, namely this freezing cold empty nothingness that occupies the whole Universe.

The only problem is that, while my mind knows that this freezing cold temperature does exist as a tangible something, I cannot see or touch it, in spite of its unquestionable existence, as making up more than 99.99% of our Universe. This freezing cold emptiness is really 100% as God's shell body, where God keeps the heated weight that ***IT*** exists as, inside of ***ITSELF***. This is what we understand to be this huge Universe, even though we may never be able to see this part of God with our physical eyes, there is no doubt that God exists as ***ITS*** freezing cold shell body, within which what we know as physical matter resides.

⌘~~~~~~~~~~~~~~~~~~~~~~⌘⌘~~~~~~~~~~~~~~~~~~~~~~⌘

*****It is ITS cold nothingness that directs where ITS heated weight may exist as ITS omnipresence ******

❁~~~~~~~~~~~~~~~~~~~~~~❁~~~~~~~~~~~~~~~~~~~~~~❁

Let us now discuss what scientists have discovered about pure energy, which is nothing other than God. They know for a fact that this pure energy exists and they know a lot about how it behaves. They know that this pure energy is what composes all the matter in the Universe; that matter is made from heated weight in the form of atoms which total 4.6% of the Universe. According to their observations, this matter exists inside a huge emptiness, also made of this pure energy which exists at a freezing cold temperature. Neither this 4.6% matter nor the huge freezing cold emptiness within which it resides cannot be created or destroyed, but only transformed. That is the way this pure energy behaves, and this transformation they have given the name transmutation. This is the process to which I have been referring in this book using the term "reshaping". Now, when we compare this heated weight in terms of the space it occupies, scientists estimate at 4.6% of the total expanse of the Universe, but I estimate to be more likely less than 1%. The reason for this is that I am factoring out the 95-98% emptiness that exists inside each atom. This would leave us with a

remainder of 99.99% emptiness, which exists at a freezing cold temperature, which is nothing other than the Universe's "shell-body", so to speak. And it is only on account of ***ITS*** heated weight that exists in the form of protons, neutrons, and electrons that my eyes are able to see this something that exists as matter.

So returning to God's existence in terms of something tangible, it is logical to conclude that ***IT*** (God) has to be made of something. And if ***IT*** exists as something, how does one see IT? If one construes God's existence as merely spiritual, then ***IT*** exists as something but not as something that exists materially. This status puts ***IT*** into existing as a nothingness or "spirit". However, the only nothingness that exists in all places at the same moment is ***ITS*** nothingness, for this nothingness is also inside of the atoms that make the 4.6% matter in the Universe possible, and this nothingness is what permeates this whole Universe as ***ITS*** shell body. Furthermore, the only place to which we can apply the meaning of Omnipresent, is ***ITS*** shell body IT, because although ***IT*** may be a shell body that exists as a nothingness, this nothingness still has one unique quality, which is that ***IT*** has a measurable distance (expansion or extension), that maintains a freezing cold temperature, that exists as an energy that cannot be created or destroyed, be ***IT*** as "God's temperature" or simply as the dual temperatures at which this pure energy which cannot be created or destroyed now exists as this whole Universe.

So returning to God's existence in terms of something tangible, it is logical to conclude that ***IT*** (God) has to be made of something. And if ***IT*** exists as something, how does one see ***IT***? If one construes God's existence as merely spiritual, then ***IT*** exists as something but not as something that exists materially. This status puts ***IT*** into existing as a nothingness or "spirit". However, the only nothingness that exists in all places at the same moment is ***ITS*** nothingness, for this nothingness is also inside of the atoms that make the 4.6% matter in the Universe possible, and this nothingness is what permeates this whole Universe as ***ITS*** shell body. Furthermore, the only place to which we can apply the meaning of Omnipresent, is ***ITS*** shell body, because although ***IT***

may be a shell body that exists as a nothingness, this nothingness still has one unique quality, which is that it has a measurable distance (expansion or extension), that maintains a freezing cold temperature, that exists as an energy that cannot be created or destroyed, be it as “God’s temperature” or simply as the temperature at which the pure energy of the Universe exists, which as energy cannot be created or destroyed.

Now here is something to think about, which is if the freezing cold Universe in terms of measurable distance could be destroyed, where would the measurable distance that now exists as this freezing cold Universe go? Furthermore, if the heat that now exists were destroyed, where would this heat go? I am sure of one thing: If all the heat were destroyed, this Universe would be a freezing cold place! And if the freezing cold were destroyed, then where would this heat exist? So I must again thank ***IT***: Thank you for allowing us to exist as your dual, extreme temperatures as one discrete moment of your existence.

Now I’d like to discuss information that scientists have gathered concerning how hot and how cold ***ITS*** dual temperatures are.

The hottest temperature we've ever been able to know of is 100 million million degrees, which was achieved in the Fermi laboratory located in the American Midwest, where they briefly reached a temperature of 10^{14} °C (100 million million degrees). According to scientists, the lowest possible temperature that can exist is what is known as “absolute zero” at -273.15 °C or 0° Kelvin or K. (named after the physicist, Lord Kelvin). Absolute zero is that temperature where atoms cease to vibrate. So far the lowest temperature achieved has been 0.0000000001 degrees above absolute zero, at the Low Temperature laboratory at University of Helsinki.

Now I would like to reflect on ***ITS*** two temperatures as ***ITS*** primary way of existence. Going back to the moment of the Big Bang, what existed then was ***ITS*** two temperatures: first as ***ITS*** + 1000 degree heat, in the form of dense heated weight, and second, ***ITS*** -100 degree cold temperature , for these two temperatures cannot be created or destroyed, for this is the way ***IT*** exists, as

what we know exists as pure energy .

Should you, the reader, be wondering why I have given such importance to something like temperatures, it is because, in reducing the pure energy that this Universe exists as, we will find that two extreme temperatures exist and since we are dealing with pure energy, they cannot be created or destroyed, and on the other hand, we can empirically verify that these two temperatures do exist. For example, our Sun's core temperature has been measured at 27 million degrees Fahrenheit. On the other hand the coldest temperature scientists have found is 0.002 degrees above Absolute Zero (-273.15 ºKelvin) which is that of the cosmic microwave background radiation left over from the Big Bang.

Now, in order to understand our existence as this pure energy, keeping in mind everything that we have said before about ***ITS*** temperatures and indestructibility, scientists have also found that this pure energy has the property of being able to transform or transmute (reshape, as I say). In other words, everything that exists in this Universe, down to the smallest sub-atomic particle, including the Universe ***ITSELF*** is nothing but this pure energy, which human religious minds call God, and ***IT*** is in constant transformation or reshaping. This is what the human mind understands as the God that created everything that exists and is in all places at the same moment; omnipresent.

There is no contradiction between this last statement and what scientists have discovered about pure energy, and this is why I welcome whatever input scientific minds can offer us concerning the way this pure energy existing as these two extreme temperatures behaves. There is no contradiction between this and the way most of us in general see or understand God, as the Creator of everything, for that means that ***IT*** too as God created these two extreme temperatures, the heat and the freezing cold that our Universe exhibits. This also includes all the celestial bodies that exist in this place that is nothing but God's being, for God is in all places at the same moment.

I am particularly grateful that I can use both of these views, the scientific and the spiritual, for this way I can understand ***IT*** better, be it in terms of the information that science has to offer about ***IT*** as pure energy or in terms of how I viewed God from my existence when I was down and out on the streets, and finally, I must never for get to also include the way I have inwardly experienced ***IT*** through meditation.

So here I must insert a thank you note: Thank you (to ***IT***) for letting me understand you as you exist in me and as how I exist out here with my limited view of how you exist as these two temperatures, and for all the information that science has amassed concerning you as the Universe, that goes as far back as the moment of the Big Bang, all the way up to this present moment, as you now are existing. This is a day and age when we are able to see and understand some of the things you do with your heated weight, that can only exist inside of your freezing coldness, where you are using black holes to return again to a new beginning where you will again use your heated weight, that exists within your freezing cold body to become what only you know.

I'd like to clarify why I am dealing with ***ITS*** weight at the moment of the Big Bang. The reason for this is that I feel that anything that can exist in the stage we are in now could only come from the way ***IT*** existed at the moment of the Big Bang, in terms of ***ITS*** temperature at +1000 degrees. This is when ***IT*** had all ***ITS*** weight in one place at a temperature in the billions of degrees rather than the stage we are in now, which began when ***IT*** fragmented ***ITS*** weight into uncountable portions which are now surrounded by more of ***ITS*** surface area that exists at ***ITS*** freezing cold temperature.

I know that there are readers who know a lot more about this subject of ***ITS*** temperatures and if there is anyone who has information that can help us understand ***IT*** better, please e-mail it to me so that I can post it on my web page's newsletter.

To continue, I'd like to give my readers another reason why I have given this issue of temperature so much attention. To understand, it is important that we remember that humanity as such has not been around for a long time, but ***IT*** as pure energy, as God, has been here forever. Keeping that in mind, if we mentally try to go as far back as possible with the information that we have gathered during our human existence, we arrive at the point when ***IT*** was existing just before the Big Bang, and all that existed then was ***ITS*** very hot, dense weight, within ***ITS*** very cold nothingness. There you have it: ***ITS*** two extreme temperatures! Knowing this we are now able to understand and, indeed, witness, how ***IT*** prepares to bring back this part of ***ITSELF***, as heat, using what we understand as black holes, to prepare what will be the next big bang. However, we, as humanity, will only be able to witness this procedure that ***IT*** is using up until a certain point, for our own galaxy will eventually be consumed by a black hole.

Then again ***IT*** might just spare us, by placing us far enough away, along with enough of the things we would need in order to survive. This way our long journey as humans in this galaxy, which started with ***ITS*** reshaping from when we were cave dwellers on this planet up to the new humans called astronauts, that have the possibility of producing a newer generation of children born on space stations. I guess that when this happens, they will no longer be earthlings, and they then might be called humans born in ***ITS*** clear, freezing cold nothingness. Now this would also mean that we would have to keep track of the nearest black holes, because we would have to have space stations that could be moved far enough away from the last black hole.

And this would also mean that the last black hole would not be totally comprised of ***ITS*** heated weight in the form of that last black hole, or as a new big bang, for we would have to deduct the heated weight of those humans and the space stations, along with all the necessary equipment and food and water for survival that would have to take us through ***ITS*** new reshaping of ***ITS*** heated weight back to a new galaxy where we could find a new sun, that would supply us with the energy we need. However, should this

happen, there would also be a difference in terms of the beginning of a new human society, compared to when we came into existence on Earth with no clothing and no scientific knowledge. This new arrival would be quite different, with those humans being very well equipped in terms of gear and also, possessing a very advanced technology. I hope that by then that humanity will have been able to overcome racism, having learned that we all come from ***IT*** as ***ITS*** searches for all existing possibilities as ***ITSELF.***

It seems to me that at the moment when the last black hole is consumed and is about to become thrown outwards as ***ITS*** heated weight, as a new big bang, whoever may be living on these space stations will also have to stay as far away as possible from this explosion. Otherwise, they risk being hit by ***ITS*** weight that is moving outwards. It is clear they will have two problems: one, not being pulled into the last black hole, and the other, not being hit by this same heated weight that will now be moving outwards. Who knows? If this ever happens I hope these space stations are equipped to record and store ***ITS*** historical transformation from big bang #1 to big bang #2. If we fill in more details of the many things that will happen during this possible stage of our human existence we might just be able to produce a film called *The Latter Day Humans That Escaped the Last Black Hole!*

So now I'd like to share a possible scenario: It is the year 3000 and since the scientific community on Earth has already picked up signs that there is a black hole within our galaxy, we already know that our entire solar system and indeed our whole galaxy will be consumed by this nearby black hole. For this reason, humanity has been aware for some time that it would have to leave its earthbound existence in search of survival.

Humanity has begun a mass exodus on spaceships and all the passengers are aware that they have no return tickets. They are also aware that as they travel part of the crew will always have to be in hibernation. Doubtless, they will mate among themselves in a way to genetically produce the best future crew in order to continue this

unknown venture. They will have accepted that the races must cross breed so that there will be as much genetic diversity as possible.

They have all been vaccinated against all infectious diseases and are as careful as possible so as not to carry any disease into outer space. They have plenty of entertainment and there is always a game of chess, or checkers, or backgammon, and they have plenty of movies that were pre-recorded before departure to films that they knew that they had not seen yet.

They are continuously alert so as to not come too close to any black holes, and as soon as one appears, they move their craft in the opposite direction, for they have mastered the art of using the nothingness that exists as the empty Universe which has no resistance to ***ITS*** heated weight, so that the heated weight that they exist as can move freely. They have discovered that once you are moving at a certain speed in this nothingness one can continue traveling at this speed forever, unless one encounters a material obstacle.

They have equipment that can manufacture other equipment that maybe necessary to further their journey within the huge nothingness that this Universe exists as pure energy (or as God, as omnipresent).

And here is the strangest thing that relates to time for these travelers. As they travel through the vast nothingness that comprises the Universe that is approaching the stage when there will be only one black hole left, they will find that there is no distance to be measured for there will be no material reference points other than the developing singularity, for these new space travelers know that as soon as all black holes are united into this last one, they will still be traveling within ***ITS*** timeless nothingness, even though they will not have to measure distances from one portion of ***ITS*** weight to another.

As the last two black holes merge into one, they find that they have been able to avoid being destroyed and they will be the first human beings to see what takes place at the moment of the new big bang. This may be difficult because ***ITS*** heated weight is also clear and transparent as this heated mass, but their scientific instruments would most certainly be able to detect it because it is was at the circumference of the singularity where ***ITS*** duality made contact, such as the contact that occurs when a positive polarity and a negative polarity come together since this is what gave rise to light which is made up of all the colors in the spectrum.

And as they watch the formation of this new matter into new galaxies, they notice that one of these galaxies appears to contain a solar system with a planet that has water and all the necessary requirements for their survival.

After they send out several probes to explore this new planet, they determine that it is suitable and they proceed to land in large numbers. Of course this planet could have different characteristics from Earth:. Its ratio of water to land may be different. The rate of its rotation may be slower or faster as well as how long it takes to go around its sun. Humanity will have to make adjustments because their mother planet's mechanical time system will no longer be applicable.

⌘~~~~~~~~~~~~~~~~~~~⌘⌘~~~~~~~~~~~~~~~~~~~~~⌘

** ***Even sex and birth are based on ITS heated weight*** **

❀~~~~~~~~~~~~~~~~~~❀~~~~~~~~~~~~~~~~~~~~❀

I am sure that this new generation of humans will be aware of how ***IT*** exists as omnipresent and as every atom that makes up their physical bodies. They will fully understand that ***IT*** is not creating anything new. ***IT*** is just reshaping ***ITS*** heated weight and that every thing that exists or will exist as life, will be alive because ***IT*** is alive, imparting mobility to everything. They will feel secure and at peace because they will know nothing is ever created or destroyed, for it is all ***IT.***

I consider this a hopeful situation for people who think more about humanity than about ***IT***. I also think we would gain a better

perspective if we remember that ***IT*** never stops or dies, and that ***IT*** is always looking for other possibilities into which ***IT*** can reshape ***ITS*** heated weight within ***ITS*** freezing cold shell body. Let me also mention, that in understanding ***IT*** better, and by this I mean being fully aware that all ***IT*** is doing is just reshaping ***ITS*** heated weight within ***ITSELF*** as ***IT*** looks for all existing possibilities, we might discover that ***IT*** has already reshaped ***ITSELF*** as the genes required to form our human bodies before the Big Bang that began the Universe as we presently know it. Perhaps we are just the latest version of ***IT*** reshaping ***ITSELF.***

But then again whether ***IT*** has, or has not done any of the above, it really makes no difference. Thinking about these things only helps me to understand ***IT*** better and what is of true importance to me is that ***IT*** gave me a chance to exist as one moment of ***ITS*** existence in this stage of ***ITS*** reshaping ***ITS*** heated weight.

It is easy to understand this, if we remember that ***IT*** has always been here, and will always be here, as what now exists as what we understand as the pure energy that exists as this Universe. This is precisely the reason why this pure energy cannot be created or destroyed. Look at it this way: ***IT*** has always existed as ***ITS*** freezing cold shell body, where ***IT*** is always reshaping ***ITS*** heated weight inside of ***ITSELF***, as ***IT*** entertains ***ITSELF*** with what ***IT*** can do. Now that we as humans are better educated than when we were at the more primitive stages of our existence when we still dwelt in caves, we have the capacity to understand that ***IT*** might have existed as many, many more previous big bangs, because what we call a big bang is really just the way ***IT*** continues to reshape ***ITS*** heated weight that exists within ***ITS*** freezing cold shell body, for ***ITS*** freezing cold, clear nothingness is how ***IT*** exists as a changeless constant. It is only ***ITS*** heated weight that is always, constantly changing within ***ITSELF.***

After this excursion into what might transpire in the far future of humanity, let us return to the subject of ***ITS*** temperatures, when ***ITS*** heated weight existed in one place as it did just before the moment of the Big Bang, ***ITS*** temperature in unimaginable zillions

of degrees in one single place was easier to maintain, because only the outer surface area of that singular point was exposed to ***ITS*** freezing cold body.

However, when ***IT*** fragmented ***ITS*** heated weight into portions, these fragmented portions are each fractions of ***ITS*** one total heat in zillions of degrees. So each fragment is now surrounded with the same freezing cold temperature at which ***IT*** has always existed, for this part of ***ITSELF*** as ***ITS body***, is constant when existing outside of how ***IT*** exists as atoms. You may understand this better if you recall that ***ITS*** weight as heat is only less than 1% of ***ITSELF*** compared to ***ITS*** huge, freezing cold body.

Now returning to the way ***ITS*** fragmented heat exists as the hydrogen that is the commonest element in the Universe, these fragmented portions of ***ITS*** heated weight are now surrounded by 95-99% of ***ITS*** freezing cold temperature. So now you will realize that in the same way that in any atom fragmented portions of ***ITS*** weight are present in the form of protons and neutrons surrounded by ***ITS*** oneness in the form of electrons, so this once freezing cold temperature we have been referring to as ***ITS*** -100 degree temperature that exists as the empty Universe can exhibit a different, warmer temperature in certain areas.

This is so because when ***IT*** had all ***ITS*** weight as heat, in terms of temperature, in one place, as in the moment just before the Big Bang, the outside area of that weight was the only part exposed to ***ITS*** freezing cold temperature. This would mean that there would be a difference in temperature in that area where both these temperatures come into contact and it would exhibit a gradation of intermediate temperatures.

Looking at it from the opposite point of view, since ***ITS*** body exhibits a -100 degree temperature; when ***IT*** took portions of ***ITS*** weight to become electrons, ***IT*** used these electrons to separate ***ITS*** -100 degree temperature on the outside of atoms from the inside temperature of atoms exhibited by the portions of ***ITS*** fragmented weight in the form of protons and neutrons. The easiest

way to conceive of this is imagining a single hydrogen atom, having one proton and one electron. The proton is a fragmented portion of ***ITS*** weight surrounded by 95-99% of ***ITS*** once -100 degree temperature. I use the word "once" because the temperature inside the empty part of each atom which was once a part of ***ITS*** -100 degree nothingness is now different because of the way ***IT*** separated the outside temperature from the inside temperature, which is now warmer, or at least it should be warmer because there is an area where ***ITS*** -100 degree temperature comes into contact with ***ITS*** heat that now exists as a fragment of ***ITS*** once +1000 degree temperature in proton form. So, still keeping a single hydrogen atom in mind, you can see why the electron, being made from ***ITS*** high cold speed, with an attached tiny fragment of ***ITS*** heated weight, has to exist as a dual temperature, at least temporarily.

I say temporarily, because ***IT***, as a whole, exists as two extreme temperatures, and also, because of the way ***IT*** uses both ***ITS*** temperatures to put together atoms. This is only a temporary structure because as soon as an atoms electrons are removed, the -100 degree temperature that exists outside will incorporate what existed as the warmer temperature that existed inside the atom as ***ITS*** nothingness.

Keeping this in mind, let us take a look at the Periodic Table of the Elements, starting with element #1, which is hydrogen. Every atom of hydrogen has one portion of ***ITS*** +1000 degree temperature as ***ITS*** heated weight (one proton) that is surrounded by ***ITS*** -100 degree temperature.

Element #2, which is helium, has two portions of ***ITS*** +1000 degree temperature as ***ITS*** heated weight and each one of these is surrounded by ***ITS*** -100 degree temperature; and so on. As one continues to peruse the list of atoms in the Table, the number of fragmented portions of ***ITS*** heated weight coming from ***ITS*** original +1000 degree that they hold in their nuclei increases. The Periodic Table of the Elements is a list of different possibilities that ***IT*** used to configure ***ITS*** fragmented +1000 degree weight so

that matter could exist and so that you and I could exist to confirm ***ITS*** existence.

Let us go back to the hydrogen atom. Now that this one portion of ***ITS*** +1000 degree heated weight (the proton) is surrounded by ***ITS*** -100 degree freezing cold temperature, the atom's overall temperature has decreased while the temperature of the nucleus has increased. This process is repeated over and over again as one runs through the list of elements in the Periodic Table and as an atom gains more of these portions in the form of protons and neutrons, the temperature of the nucleus will increase. This is so because, while the nucleus is still surrounded by what was once ***ITS*** -100 degree temperature, it is cut off from the totality of it by the orbiting electron(s). This means that this area of the atom will exhibit a warmer temperature since it will be in contact with the nucleus and will be absorbing its heat from the +1000 degree fragments (the protons and neutrons). The principle behind my line of reasoning is this: ***IT*** as pure energy exists in the form of dual temperatures, that is, ***ITS*** +1000 degree heated weight which dwells inside ***ITS*** -100 degree freezing cold body, and both of these temperatures, as pure energy, cannot be created or destroyed.

Here is the same thing form yet another perspective: Imagine that you are looking at ***IT*** from a distance, perceiving ***ITS*** dual temperatures as different polarities, that is, positive and negative, while at the same time remembering that ***ITS*** nothingness is basically a form of a freezing cold, clear, transparent energy. Since we have to give this area of nothingness a name, let's call it negative energy, since we have already labeled ***ITS*** heated weight in the form of protons as positive energy. I also consider ***ITS*** weight in the form of protons and neutrons to be transparent because when I look at an object from a distance I know that I am seeing it through billions of portions of ***ITS*** fragmented weight in the form of protons and neutrons that compose the elements that make up our atmosphere. This is also the case when one looks through an Earth based telescope, and when one looks through a telescope in orbit around the Earth, such as the Hubble Telescope, there is no atmosphere, but whatever one view or photographs

must be seen through an unimaginable number of hydrogen atoms which constitute 90% of the Universe.

Continuing with our thought experiment, if we were looking at ***IT*** from a distance as ***IT*** existed at the moment just before the Big Bang when ***IT*** had all ***ITS*** heated weight concentrated into a single point somewhere within ***ITS*** freezing cold, transparent, negative nothingness, we would only be able to see the glow of light that would exist only around the circumference of ***ITS*** heated weight. This light would exist because it would be a product of the contact between ***ITS*** heated weight, as a positive source of energy with ***ITS*** freezing cold nothingness, as a negative source of energy. It is this interaction between positive and negative energies that gives rise to light which is made up of the entire spectrum of colors.

Now, this light came from ***ITS*** freezing cold, negative energy making contact with ***ITS*** heated weight as a positive source of energy at the border of the circumference of the singularity only, for IT, as well as light, is a duality in which two energies exist together at two different speeds: ***ITS*** freezing cold speed that is even faster than 186 thousand miles per second (which I refer to as ***MAXX-SPEED***, to which I will arbitrarily assign a speed of 200,000 miles per second, a number for the sake of this discussion only) and a lower speed which we call the speed of light (186,000 miles per second). ***ITS MAXX-SPEED*** is that part of ***IT*** that exists as ***ITS*** nothingness. From this nothingness a fraction of ***ITS*** heated weight came into existence in the form of light, which is what ***IT*** uses to move ***ITS*** heated weight from one place to another within ***ITSELF*** at 186,000 miles per second. The next step was the formation of the first hydrogen atom, which ***IT*** accomplished by using more of ***ITS*** heated weight and slowing down the speed in order to form a proton and an electron. From this, the next step the formation of all the naturally occurring elements listed in the Periodic Table, which gave rise to matter. This is the way ***IT*** used ***ITS*** dual temperatures to become matter, so that you and I could confirm ***ITS*** existence.

For those readers who practice meditation, using what is referred to as the Third Eye, what you perceive as the Third Eye has a glow of light around its circumference, where there is really nothing inside or outside. This is similar to the way ***IT*** existed just before the moment of the Big Bang. On the other hand, please do not spend time trying to analyze the Third Eye because this will keep you from making and enjoying direct contact with ***IT*** as ***ITS*** nothingness. Furthermore, I need to remind the reader that the information I have offered comes from my interpretation, from my intuition and insight concerning how ***IT*** exists as ***ITS*** two extreme temperatures. So, I have tried to explain how ***IT*** can exist as ***ITS*** outside and ***ITS*** inside simultaneously, at a temperature that we have registered to be -100 degrees as ***ITS*** shell body, where this something (***IT*** or Pure Energy) keeps ***ITS*** heated weight. This is what the scientific community has postulated about how ***IT*** existed just before the moment of the Big Bang as a singularity where ***ITS*** weight as density was in the tons per square inch having a temperature in the trillions of degrees as ***ITS*** heated weight.

Let me add that in my search to understand ***IT*** better as ***IT*** exists as omnipresent, I can only say that the more I have found concerning the ways ***IT*** exists as hot and cold, as weight, and as nothingness, I have had to accept that ***IT*** is incredible in the things ***IT*** does, and as the strange ways that ***IT*** exists, as what is known in science as pure energy or to see what ***IT*** does with ***ITS*** heated weight as ***IT*** reshapes. There is no greater satisfaction than my relationship between the way ***IT*** exists as me, and the way ***IT*** has permitted me to directly connect with ***IT*** through meditation, as ***IT*** exists inside of me, and the way ***IT*** interacts with me and as what ***IT*** exists outside of my human body, as the things in which ***IT*** presents ***ITSELF*** to me, in the form of people and events, as ***IT*** is reshaping ***ITSELF*** as this show that we are seeing. I am grateful that I have been permitted to observe while ***IT*** allows me to exist as just one moment in ***ITS*** existence as a gift called "my life", for I can personally say that I do not get the same feeling when I merely use my intellect to think about what ***IT*** may be doing with ***ITS*** heated weight somewhere else inside of ***ITS*** freezing cold, nothingness shell body, for I know that my gift is not there as ***ITS***

weight. ***IT*** is here, wherever I am that I can be directly with ***IT*** as ***ITS*** heated weight in the form I presently exist and as the 95-99% nothingness that is also a part of my existence as ***IT.***

Here's another helpful analogy to understand this God that exists as pure energy: Imagine you are God and that you have a body that is freezing cold, and inside of this cold, transparent body of yours, you can have all your heated weight in one place, (as was the case in the moment just before the Big Bang occurred), or that you can fragment your heated weight and reshape it into other possibilities or other ways of existing, knowing that no matter what you do with your heated weight, there is no possibility of losing any of it because it cannot leave your freezing cold shell body. This is an accurate description of our Universe.

God in a nutshell

I hope no one is offended at my using the phrase "in a nutshell" about God. What I want to convey is putting something into words in the briefest, most concise way possible. Of course it is impossible to fit ***IT*** into a nutshell!

So this is the shortest possible description concerning God's existence: All that God is doing is moving ***ITS*** inner heated weight around inside ***ITS*** freezing cold shell body as ***IT*** searches for all the possibilities for ***ITS*** existence, in terms of ***ITS*** heated weight. And when ***IT*** moved ***ITS*** heated weight at ***ITS MAXX SPEED*** we got the phrase "and there was light".

⌘~~~~~~~~~~~~~⌘~~~~~~~~~~~~~⌘~~~~~~~~~~~~~~~⌘

**** ***O God: Thank you for allowing us to 1- confirm that you exist, and 2 for allowing us to participate in moving your heated weight within your freezing cold, clear nothingness*** ****

۞~~~~~~~~~~~~~~~~~۞~~~~~~~~~~~~~~~~~۞

Part #2
OMNIPRESENT AS TIME

What is time?

There is the human time zone that we understand as our mechanical time system. This time system is only a human convenience. We use clocks that follow a 24-hour cycle because our planet spins at a rate of 24 hours per rotation. We call this 24-hour cycle one day. We use calendars that follow a 28 to 31 day cycle because of the Moon. It takes the moon 28 to 31 days to rotate around Earth. We call each rotation of our Moon around the Earth one month. We base our year upon the rotation of the Earth around the Sun, which takes the earth 365.25 days to complete. Mathematically, if one day equals one earthly rotation, then one year equals 365.25 rotations. Measuring a year by the Earth's rotation around the Sun isn't applicable to the rest of the Universe, however. A year marked by 365.25 rotations is what we as humans have established as one Earth year.

❁~~~~~~❁

We can exist without time, but time cannot exist without our presence

⌘~~~~~~~~~~~~~~⌘~~~~~~~~~~~~~~⌘~~~~~~~~~~~~~~⌘

If you are 50 Earth years old you could say that you have lived through 18,262 Earth rotations. If you are 30 years old, you have been here for 10,957 rotations. Multiply your age by 356.25 (number of rotations per year) and that will determine your age in rotations. Should you die when you are 100, you will have been here 100 x 365.25 days, which is 36,525 earth rotations. These rotations have permitted you to do all the things that you did while you were here on this planet.

It is a marvel when we say Jesus Christ was here 2,000 years ago. Two thousand years multiplied by 365.25 rotations for every year would equal 730,500 rotations. So, 730,500 rotations ago Jesus Christ walked the Earth. It is the rotation in reverse that would put Jesus Christ here again, for rotation is a real physical event.

To explain the other time zone, the place of omnipresence, let's return to the example of the thirty year old: If you were to go backwards by 10,957 rotations, you would have just been born. If

we were to look at this in slow motion, we would see you being born, and as you developed we would see no change. Change happens moment by moment. This moment-by-moment change took place for 10,950 rotations, which resulted in you as you are at this moment.

Humans start as the union of an egg and a sperm that are so small they are microscopic. We are born and start out as one moment of existence. We change and develop as one moment that is searching for a maximum weight. After reaching our maximum we begin to collapse or die. This makes it possible for ***IT*** to reshape again, to transmute.

It is not that one moment follows another moment, but rather the same moment reshapes or "changes" as the same omnipresent moment.

Have you ever seen a high-speed film of a seed growing into a flower and then withering and dying? If you have, what you have seen is the absence of the effect of Earth's rotation. This high-speed film allows us to see how things change as one moment, as the flower reshapes itself on the basis of the same existing moment, as a place for this birth, development, and dying process to occur.

Returning now to our concept of one day, we say "day" (a period of solar light) because we talk, think, and count in days. We keep track of the things we do during our days because this is what we call time. The majority of us seldom count the nights because that is when we sleep. Due to this, nights do not have much importance in our lives as far as they relate to time.

*****We do not exist because of time, time exists because of us*****

When humans first roamed this planet, there was no need to count days. Some people used the Moon for keeping time, hence the phrase, "many moons ago." It is our civilization that developed the human mechanical clock or watch. We now live by this development, by a twenty-four hour day that consists of a duality

of day as light and night as darkness. The Benedictine monks are credited with giving us our way of using time. They began the day with a morning prayer.

The Industrial Revolution increased the use of this human time system. People needed to be at the same place at the same moment (time) for factories to operate.

For the majority of us, a week consists of Monday through Sunday; this too is a matter of human convenience. The Monday through Friday week exists so that we can work and transfer our human energy. We attempt to make Fridays and Saturdays last as long as possible by staying awake longer. We traditionally dedicate our Sunday mornings to our God, and Sunday afternoons are used to readjust and prepare for Monday.

❁~~~~~~~~~~~~~~~~~~~~~~~~~~~~~~~~~~~❁

****** *IT has always existed timelessly.* *****

⌘~~~~~~~~~~~~~~~~~~~~~~~~~~~~~~~~~~~⌘

Using the term 2005 as a label for a year in time is also a convenience. When we celebrate the moment we call New Year, other parts of the world are not in the New Year as the same moment.

Today this system governs our lives. We know time by reading mechanical clocks that divide our days into hours, minutes, and seconds. Our lives revolve around this human mechanical clock because it is a very efficient system. This system provides for people to be where they need to be at a specific moment. This human mechanical time system works well on Earth. It was designed by the human mind for the things humans do on this planet.

❁~~~~~~❁

***** *PURE ENERGY EXISTS AS A TIMELESS PLACE******

⌘~~~~~~~~~~~~~~⌘~~~~~~~~~~~~~~⌘~~~~~~~~~~~~~~⌘

I call our planetary time "solar rotational biological time," for it is based on a combination of solar energy, planetary rotation, and biology. Time is solar and biological because it is the Sun that gives us the energy we need for everything that functions

biologically, and time is rotational because it is the rotation of our planet that gives us the illusion of day and night.

Earth's rotation also helps generate the gravity that makes it possible for us to stay on its surface. Rotation gives us the feeling of change. It causes the confusion in our minds as to what we understand as time. We see things change because of this rotation, this spinning. The mind instituted time as we know it as a way to establish its dominion and control.

The 24-hour day is easier to understand and is most applicable near the Earth's Equator. North of the Equator, the concepts of time, day and night change somewhat. Above the Artic Circle in the North and in Antarctica at the South, there is continuous sunlight for 6 months. The remaining period of 6 months is spent in the dark. I would think that for people living in those areas, terms such as "one day follows another" and "tomorrow will be another day" have a slightly different connotation or feeling than they do for us closer to the Equator.

⌘~~~~~~~~~~~~~~⌘~~~~~~~~~~~~~~⌘

******Time cannot exist without motion******

❀~~~~~~~~~~~~~~~~~~~~~~~~~~~~❀

Besides location, situations also affect our sense of time. When we are separated from someone with whom we are madly in love or infatuated, an hour can seem like an eternity. On the other hand, when we are together and enjoying ourselves, time flies. We may also feel an absence of time when we meet an old friend, as though we saw him or her only yesterday when it may have been 10, 20, or 30 years since last we saw each other. We say it feels just like yesterday because our mind believes that because we had slept prior to the encounter, the time elapsed had been at minimum a day, hence yesterday.

❀~~~~~~❀

****Jet lag occurs when the mind and body have to re-adjust for crossing different time cycles, as the same existing moment.****

⌘~~~~~~~~~~~~~~⌘~~~~~~~~~~~~~~⌘~~~~~~~~~~~~~~⌘

In reality, what occurred 10 years ago happened 3,652 Earth rotations ago, but it also happened in this same moment as a place

known as omnipresent.

Imagine that you are watching a movie on your VCR at night and your eyes get tired. You decide to close your eyes and rest them for a moment. Once you have rested your eyes you open them again and continue watching your movie. You are still in the same living existing moment, and in the same place as the same omnipresent moment, as before you closed your eyes.

❁~~~~~~❁

***** *Here's something else to think about: Time did not exist before the Big Bang. But IT did and still does as omnipresent. If IT does not exist as time, why do we want to impose our mechanical time on IT? IT does not need it.******

⌘~~~~~~~~~~~~~⌘~~~~~~~~~~~~~⌘~~~~~~~~~~~~~⌘

You are still in the same moment-as-a-place when you where given this gift called life, and it will be the same moment where you will die. We believe that we are constantly moving forward. What we know as the past is what was left behind. When the Sun goes down, that day is in the past. Everyone believes this; it is natural to understand time in this manner, for it is comfortable.

During the age of primitive man, there was no way to understand time as anything other than a moment of existence. We use the planet's rotation to count millions of years from the beginning of Earth's existence. These numbers sound extreme to us, but ***IT*** measures in these types of numbers. To us a billion is a large number, but by the time you are 20 years old your heart has pumped more than a billion times. ***IT*** does ***ITS*** reshaping in a repetitive manner. Although things are repetitive, each repetition is different because something has been reshaped. All this occurs as the same moment. Right now is still the same moment in which you were conceived. As you read this, it is the same place that the Big Bang occurred. It is not the moment that is changing. What we call a moment is actually a place wherein ***IT*** continuously reshapes.

❁~~~~~~~~~~~~~~~~~~~~~~~❁❁~~~~~~~~~~~~~~~~~~~~~~~❁

***** As pure energy, IT reshapes and has no need for our mechanical time system *****

As we reshape into old age, we, along with this pure energy, work as a duality. ***IT*** reshapes into what we see outside of ourselves and into what we are inside of ourselves.

We, as a human body, come from nothing; we see our body going from our beginning into old age. The human body continues to reshape in search of reaching its maximum potential. It will continue to reshape until its death, at which time the energy will be transmuted into something else.

❁~~~~~~~~~~~~~~~~~~~~~❁❁~~~~~~~~~~~~~~~~~~~~~❁

**** Everything is a place where events are taking place as omnipresent.***

⌘~~~~~~~~~~~~~⌘~~~~~~~~~~~~~⌘~~~~~~~~~~~~~⌘

To exist as the existing moment, the best rule to follow is simply to do that which only you can do as that moment.

There is no yesterday or tomorrow. That is why we cannot see "into" yesterday or look "into what we call tomorrow." **Everything is here as the here and now, known as the omnipresent.**

There was a moment, as a place that existed as this Universe, before there was time. That was the moment that gave us time. This is why it would be more correct to say that matter started at the moment of the Big Bang, as a place. The dense matter that is the origin of all, existed as this place that we call the existing moment, not as the time that we are familiar with. It is not a new day. It is a continuation of the existing moment as a place.

❁~~~~~~~~~~~~~~~~~~~~~❁❁~~~~~~~~~~~~~~~~~~~~~❁

****** Everything that has happened has happened as this omnipresent moment ******

⌘~~~~~~~~~~~~~⌘~~~~~~~~~~~~~⌘~~~~~~~~~~~~~⌘

New Year's Eve

Here is something to think about, and is related to time as the moment that we call New Year's Eve. The main point here is for you to see that time is something that we have put together as a human convenience.

So that you will see that we all exist as the same living existing moment (omnipresent), and you will agree with me in what I will now say, let us start with the moment our planet makes a particular turn and is facing the Sun, which is where we say a day starts. We will use Australia as the first place on this planet that will be facing this point in our planet that we call New Year's Eve. At that moment in Australia people have agreed that most of them will stay awake so that when their clocks strike 12:00 AM there, that is the first place on this planet that will see the New Year.

❀~~~~~~❀

****REWARD******

For anyone that can hold on to the past year or New Year's Day, or grab onto and show me this old or new year.

⌘~~~~~~~~~~~~~⌘~~~~~~~~~~~~~~~⌘~~~~~~~~~~~~~⌘

Now I have to stay with our existence so that you can understand time as a convenience. I will use our first basic necessity, which is the gift of breathing and talk about the first people on this planet to reach the New Year. I want you the reader to focus on their breathing, and it is better if I use the name I gave my son which is ***Zii***. Zii takes a breath at the moment that the New Year arrives. We then go over to the other side of the planet, where we will find a person which we will call ***Delta Premie,*** **which is my daughters' name**, who is still sleeping, for she is trying to rest well so that she can have plenty of energy when she awakes, as she gets ready to receive the New Year. She needs our planet to continue turning so that when her side of the planet turns towards the Sun she will be in her New Year.

❀~~~~~~~~~~~~~~~~~~~~~❀ ❀~~~~~~~~~~~~~~~~~~❀

*** ***God does not exist as Monday through Sunday, IT exists timelessly.*** ***

⌘~~~~~~~~~~~~~⌘~~~~~~~~~~~~~~~⌘~~~~~~~~~~~~~⌘

And here again I would like you, the reader, to see that the two facts that contribute to the illusion of time are light and sleep. As the light that gives us the illusion of a new day, and as in every time we wake up from our sleep, we believe that everything previous to this awakening exists as a past, and when you become aware that all of these effects are due to ***ITS*** heated weight as matter that is changing, you will then see that all of us have always

existed inside of ***IT*** as a place which we call or know as omnipresent.

Now that you are aware of **Delta**'s existence let us go back to Zii, who is about to start his New Year. As the clock strikes 1200 AM for Zii, as he breathes in his New Year moment it is the same moment in omnipresent for **Delta**, who is still sleeping on the other side of the planet and does not want to be disturbed. She wants to feel that 12 hours later she will be in a New Year also.

❀~~~~~~~~~~~~~~~~~~~~~~~~❀ ❀~~~~~~~~~~~~~~~~~~~~❀

***** Using our mechanical time system we cannot all be in the New Year as the same moment *****

⌘~~~~~~~~~~~~~~⌘~~~~~~~~~~~~~~~~~⌘~~~~~~~~~~~~~~⌘

I have used this example so that you can see why **time is only a human invention as a human convenience**. We cannot all enjoy the same New Year at the same moment because it is based on our mechanical time system. But we can all enjoy the same living, existing moment known as omnipresent even if you are somewhere else in the Universe. We all exist as the same moment known as omnipresent, no matter what time system we are using, for we can only exist in the same moment that ***IT*** exists.

As for the accuracy of the year, we should remember that every year is off because of our need to continually adjust our calendar so that it will be synchronized with the seasons. The leap year, another human convenience, was invented to fix this problem. If this procedure were not followed we would have to add approximately six hours to each year instead, which would really confuse the human mind. So it was decided that it would be better to add one full 24 hour day every 4 years.

❀~~~~~~~~~~~~~~~~~~~~~~~~❀ ❀~~~~~~~~~~~~~~~~~~~~❀

****** There is a feeling that comes from being trapped, when you can't move forward or backward in a crowd, such as in Times Square during New Year's Eve. While arm-to-arm in a crowd, you move from the past into the New Year. It may even be freezing cold, but because of the crowd, you may not be able to leave. Imagine having to use a restroom ! ******

⌘~~~~~~~~~~~~~~⌘~~~~~~~~~~~~~~~~~⌘~~~~~~~~~~~~~~⌘

The names we gave them

The names for the days of the week in English are based on the names of our solar system's planets and on the old Germanic gods and goddesses.

The most obvious are Saturday, Sunday, and Monday; Saturn's day, the Sun's day, and the Moon's day, respectively. Tuesday is for the god Tiw; Wednesday is for the god Woden; Thursday is for the god Thor and Friday is for the goddess Frig.

According to what people believed in Antiquity the seven day week arose from the idea that each day is governed by a celestial body, namely: the Sun for Sunday, the Moon for Monday, Mars for Tuesday, Mercury for Wednesday, Jupiter for Thursday, Venus for Friday, and Saturn for Saturday.

Now let us review our months. The word month refers to the cycle of the Moon (moonth). January and March are named for two old Roman gods: Janus (god of beginnings and endings) and Mars (god of War) The remaining months' names also come from Latin: February's name comes from the old Italian god Februu or from this god's rites of purification called *februa*. Scholars think April comes from the Roman word *aperire*, "to open" on account of spring. May comes from Maiesta, the Roman goddess of honor and reverence. June was obviously named for the goddess from Juno but some scholars think it could also come from the Latin term *iuniores*, which means juniors or young men. July used to be named *Quintilis,* because it was the fifth month of the Roman year, but it was changed to July in honor of Juliuis Caesar who was born in that month. Then comes August, in honor of Augustus Caesar, the first Roman emperor. The remaining months are named for their order in the Roman calendar: September from *septem* (seven); October: from *octo* (eight); November: from *novem* (nine); and December: from *decem* (ten).

The Egyptians noticed that the Earth would take about 360 days on a full rotation. Now the full circle is credited with 360 degrees. The ancients knew that the true calendar year was longer than 360 days

and that Earth's rotation was not a perfect circle, but the mathematical circle we use in our calculations has kept the 360 degrees, because that never changes.

The original zodiac was developed by the Mesopotamians. The Greeks inherited it from them and the Egyptians.

Ptolemy gave us the Earth centered arrangement of our planetary system. He was wrong, of course, but it served us for many years. Copernicus was able to calculate that the Sun was the center of our planetary system, and not the Earth. Galileo expounded on this new model and was made to suffer for it by the Catholic Church.

He was forced to recant, to deny that this was so, but as he said himself, it did not change the fact that the Sun was the center.

All of this reflects mankind's need to find some sort of godly plan and purpose in the cosmos, a need to make sense and order of what could possibly be just an absolute drifting chaos as ***IT***.

Our solar time

If we were to encounter space travelers, they would not be using our 24-hour time system. The mechanical time system we use is based on the Earth's rotation. If it were to spin 50% slower it would change the way we measure time. If our planet made a full rotation in 48 hours instead of 24, our clocks would change from a 24-hour day to a 48-hour day. This would give the perception that we have more time in numbers only. We would not actually have more time; we would merely have longer days in numbers. If you were 50 years old with the current rotation, you would be 25 years old with the new rotation. But, you would be younger in numbers only. Biologically, you would still have lived in the same moment. The same would apply, but in the opposite manner. If the planet were spinning twice as fast, would we die sooner?

Real time is 186,000 mps

Most of us have heard the expression "real time" used when related to high-speed communications—virtual meetings or

language translations might be in "real time"; multimedia transmitted on computers might be in "real time." When we send or receive a text, sound, or image of something or someone hundreds or thousands of miles away, we call it "real time" when we receive at the same time it is being sent, such as happens in face-to-face conversation. We can now experience real time using technology. During the same moment that someone is talking to us, we can see and hear that person on our communication's device—on our monitors and through our speakers. During the early stages of this technology, we could receive only text real time. Then came sound and images, and then text was replaced with sound, pictures, and motion. It was slow motion initially, but now we can see with sound and motion that has gotten so fast that we experience it as happening in real time, as the same moment.

This effect is possible because of ***IT*** existing at 186,000 mps. Our communications move at the speed of light or at that speed where we say time stops. And this kind of talk about real time makes sense to us because we know that the information being sent is maybe thousands of miles away. What is being received must be happening extremely fast—so fast that the space that exists in between seems eliminated. Just as matter appears real and without being 95% empty, so does transmitted communication seem to occur in what we refer to as real time.

This information is not new, however. What is new is this: The main reason the events of real time can take place is because we exist on a planet that moves slower than the speed of 186,000 mps. The only reason we can even talk about the events mentioned is because we are on this slower moving Earth and because here, whatever ***IT*** is, ***ITS*** speed exists as a duality.

We and every thing on this planet exist only because of the slower speed, and we can see and feel this slower speed naturally, which confirms that ***IT*** exists at this slower speed. However, we also can confirm that ***IT*** exists at this very high speed of 186,000 mps. And this duality works together, for we and everything that exists within this omnipresence is moving at this very high speed that ***IT***

reshaped into as electrons so that ***IT*** could reshape as ***ITSELF*** as matter within a slower speed, such as that of our planet and that of our bodies. We exist in slow motion due to the high speed.

Perhaps this explains why ***IT*** exists as empty space, as a vacuum where ***IT*** can move at this speed of 186,000 mps. As empty space, this speed has little resistance for ***IT*** to move within ***ITSELF***. Additionally, since we appear in slow speed, but exist at high speed, and we appear as matter but exist as 95% empty space, and since it is speed that brings these dualities together, here is another question for the mathematicians: With all the information we have about the speed of light and the distance of space inside atoms and in our galaxy, couldn't we figure out how far we are from where we are, and how far we are from the outer border of this Universe?

A million years

When we say that something has been out there in outer space for millions of years it is a natural indication that what ever is out there came before our mechanical time had any meaning and it is out there as ***IT***.

A nice day

Because we have returned from sleep, which is how we rest, and after rest we should be feeling "good", this is where I found a sensible meaning to the word good. As for the word day, as in "good day," maybe it makes sense if we think that the majority of us are going to transfer energy in daylight in order to see what we are doing.

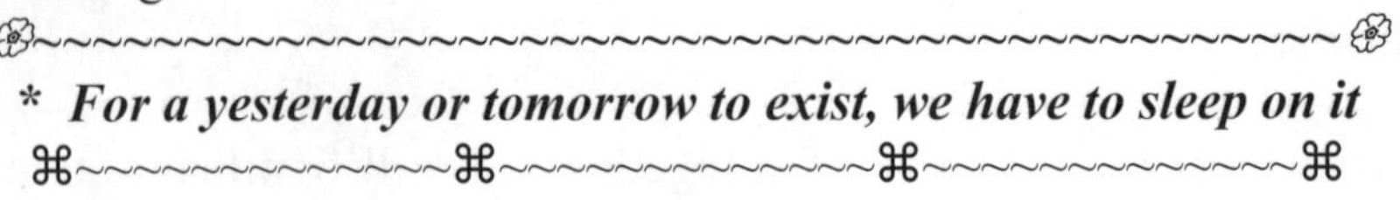

* ***For a yesterday or tomorrow to exist, we have to sleep on it***

Why there are no two days alike

Have you ever heard the saying that there are no two days alike? We say this because from the moment that we get up we subconsciously begin to notice the things that have changed, such as when you see different people on the way to work; and there

will always be something different at work, even if they're just minor changes.

We see change as something that we attach to time (a minute, and hour, a day). This is a natural thought with which we have grown up. We know now that change is not attached to time because it is just ***IT*** reshaping as the same existing moment.

❀~~~~~~~~~~~~~~~~~~~~~❀❀~~~~~~~~~~~~~~~~~~~~~❀

*******All changes are manifestations of ITS reshaping.*******

⌘~~~~~~~~~~~~~⌘~~~~~~~~~~~~~~~~⌘~~~~~~~~~~~~~⌘

A living moment

Take someone you see every day and the next time you're with that person ask him/her to remember that moment that you shared as a living moment. Ask that person to record it in his/her mind, and then look at each other, confirming your existence.

The next time you talk to that person face to face, ask if you're still in the same living, existing moment in which you last saw each other. If you really think about it you'll realize you are.

❀~~~~~~❀

******Everything that you have seen and experienced has happened in this same living moment.******

⌘~~~~~~~~~~~~~⌘~~~~~~~~~~~~~~~~⌘~~~~~~~~~~~~~⌘

Here is a different way to see the same thing: Spend a week with someone where there is no contact with the outside world, and most importantly, no contact with a clock; so mechanical time is out of your minds. Let us imagine that you can both stay awake for a full week and stay active. You will both notice that everything you do together you can only do in and as the same living existing moment.

After that week of having fun together you will find that time did not exist, because you were not affected by the rotation of the Earth or the turning of the clock.

You would have simply lived in the existing moment; time as you know it would be irrelevant. You would be independent of time. We can say that everything you did would be a continuous transfer

of energy in an existing moment that we can call the here and now, or a living, existing moment.

In reality, time is like that throughout our lives. If you remember this it will change your life.

A compressed thought

Why do we refer to time as going by fast? Let us talk about a period of time, for instance 5 years.

Let us remember that the speed in which we exist is due to the speed of our planet, which is relatively constant.

One reason we feel that these 5 years we are talking about go by fast is human memory. We should remember that any memory only truly exists in the existing moment, as a thought only.

In remembering these 5 years we are recalling as many of the events as possible. But, we are actually remembering events that took place during approximately 1,826 rotations of our planet. There will be an incredible amount of detail that will be impossible to remember. In effect, we will recall very little, and this will make us feel that those 5 years actually flew by.

To continue with time, let us take our basic day, one simple rotation of the Earth on its axis. While a rotation takes place there is 99.99% of emptiness in existence that does not change. Since ***IT*** is in constant motion, all activity causes change. Since we are made in ***ITS*** own image we take part in these changes.

Knowing this, let us now go back to our every day activities. For 95% of us, we, as kids get up and go to school, and after school we play a little. As adults we get up and get on the same roads to get to work. In many cases we do boring, repetitious work for about 5-6 rotations per week. On weekends we take part in other activities that for some of us are also boring and repetitious.

Let us look at the activities; this is important. When we use our memory to recall the many events that occurred during the previously mentioned 5 years, or 1,826 rotations, we will notice that we tend to recall not the boring repetitious activities, but the more interesting, significant events.

If we see someone in person that we have not seen in 1,826 rotations, we will recall some of the events that took place that were not repetitious, for those events were few and far-between. It is easier to separate the interesting events from the repetitious ones because we remember them as events. The mind is recalling events as things that really took place as this same living, existing moment.

The problem is that our human minds have not been accustomed to understanding that everything happens as the same existing moment. We are ingrained with the idea that all events are different because when light disappeared, it became dark, the planet made another rotation, and we needed rest. But the only thing that really happened was that ***IT*** was causing change as ***IT*** reshaped ***ITS*** weight with in ***ITS*** nothingness.

We still attach time to everything that occurs, this makes it difficult to see that everything has existed in omnipresence, independent of our concept of our human mechanical time system. What also makes it very difficult to live in the existing moment is that the majority of us must continue to work for survival in our monetary and mechanical time systems.

Why time cannot go faster

The first thing that we have to remember when we say that time goes by fast is that time is based on the rotation of our planet, which is quite constant, and that this quickness is only a human perception. In essence, it is our mind that thinks time has gone by quickly.

We as humans cannot exist outside this moment that has to take place within this omnipresence.

Think of it this way: The events that you are recalling happened because ***IT*** existed at the time of your memory as an existing moment, and the reason you can recall this memory is that ***IT*** still exists as the same existing moment.

Since the mind is not aware of this, it eliminates the boring activities which occupy much of the intervening time and finds only the desired memory.

What is an occasion?

One reason why I find that we look forward to the beginning of what we call a new day is because, since we have just gotten up from sleep (rest) we are fully charged and have plenty of energy to transfer into action.

We are ready to start something, since we are leaving our sleep, which was a stopping action. Since this planet has made one more rotation we now have light again, because light in itself is a transferring of energy. Even if what we are going to do as work is a repetitive action, it still feels different because ***IT*** has reshaped.

We know this because we listen to the news and know that things have changed. But if you look very, very closely, you will notice that you are still in the same living existing moment as that in which you were yesterday, and every moment before that.

I know you, like me, see yesterday as being a different occasion; and here is the secret: You must understand that pure energy has to reshape ***ITSELF***, and as such we tend to see ***IT*** as a totally different occasion due to our having to close our eyes during sleep, and because our planet, having made one more rotation, is now again in the same position, where light can reach us again, But, ***IT***, within all its constant change is still the same existing moment ***IT*** has always been, and always will be.

What produces the most confusion for us concerning time, is our having to sleep and our Earth spinning, for in this spinning, we get the illusion of a new day, because of the light reappearing, which

we see as the beginning of something new, and conversely, as the absence of light we get darkness to which we see as an ending, for again we will have to close our eyes to sleep, to which we have to surrender our existing moment as time coming to an end. In short **the two main ingredients for time are our having to close our eyes, and the disappearing and reappearing of light.**

Your eyes
Here is another way to see that you live as, and are a place in an existing moment. Find a room where there is no sound, the lighting is constant and there are no clocks. Imagine that you could be there and not hear anything, not feel the effect of Earth's rotation around the Sun nor feel any temperature changes. Imagine that you have everything you could possibly need for survival and you have no need to step outside of this room.

You close your eyes for a few seconds, upon opening them you are still in the exact same place and situation. You will sense that nothing has changed, that no time has passed, and that you are in the exact same moment because you have not had to relate to anything external to the moment and place you are in.
People who meditate know this feeling: You concentrate on your inner self for a long moment and find that you are in a place that has always been there while being totally disconnected from our mechanical time system.

As you are in this long moment inside of yourself, you get closer to the omnipresence of ***IT***, the true existing moment, because you exist as energy, with or without time.

Now, you might say, so what? You closed your eyes and you opened them and here you are.

But here's the thing: Remember that the closing and opening of your eyes also kept you in the same existing moment where you felt no need for time itself. Our mechanical time system is designed for us to be in a particular place at a particular moment. The moment you start moving towards a location and the moment

you arrive will remain the same moment as when you left.

In the same way, the moment you were born is the same moment you live in now. I know this is hard to digest. The biggest reason for this is that you see that your body and everything around you has changed. But we've already established that change is in the nature of ***IT***, and all that exists.
Think about this: When you go to sleep and when you wake up is the same moment of existence; it is when you start moving out of bed that you start the process of changing what is out there; which also exists as the same moment.

Time has no weight
We say that something or other happened a million or a billion years ago, but we cannot really quantify time as we can with matter. It has no atoms to give it weight, or waves or vibrations that can be measured. In other words we have not yet been able to prove that time physically, quantifiably exists, like something that has weight.

⌘~~~~~~~~~~~⌘~~~~~~~~~~~⌘~~~~~~~~~~~~~⌘~~~~~~~~~~~⌘
*** ***IT is not that the times that are changing; it is that ITS weight is reshaping*** ***
❀~~~~~~~~~~~~~~~~~~~~~~~~~~~~~~~~~~~~~❀

Motion as time
We are in the habit of saying that things started so many hours or years ago, or that something will happen in so many days or months from now; and they might indeed so start or happen, but not because we are traveling along a timeline, but because this planet will continue to make many more rotations.

Our sense of time (such as yesterday or tomorrow) results from our unawareness of Earth's spin, which is a natural event that is the result of ***IT*** putting motion on our planet, like everything else in the Universe. This is why we see what is outside of us as moving, and our mind makes sense of this movement by describing it as forward or backward, thus giving us yesterday and tomorrow.

We say that the Sun comes up and goes down, but this is just an illusion. In reality, it is the rotation of the planet that gives us this sense of another day. And since ***IT*** has continued in ***ITS*** reshaping, which we see as change, we are reinforced in our concept of there being a new and different day. The reality is that all that took place was that our planet made one more rotation and that we were unaware of ***ITS*** reshaping.

❁~~~~~~❁

****** Clocks, like time, cannot exist without motion. ******

⌘~~~~~~⌘

The Earth's rotation is not based on time, it is based on an action that ***IT*** produced in order to reshape.

I have been insisting on this subject of time, not because of time itself, but because by removing time we will be able to see ***IT*** more clearly. You must remember that ***IT*** does not exist in our mechanical time system, or as a yesterday or a tomorrow. We exist in a living moment. ***IT*** exists in omnipresence, a very dynamic moment in which we all exist.

ITS speed as a minus

You exist because you are made of atoms traveling near the speed of 186,000 miles per second. GOD cannot reshape into things like planets, stars and humans unless **IT** slows down, slower than the speed of 186,000 mps, which is where we understand that time stops.

IT slowed down to reshape into this planet called Earth. As a consequence we are moving slower than where we say "time stops." Using Earth's rotation to anchor our concept of time is mistaken, because we actually exist in a no-time zone. The Milky Way is spinning faster than Earth. It is possible that this is so because ***IT*** allows the Earth to be contained within the spinning. The Milky Way is spinning faster than Earth but slower than the speed of light, which, if we construct a number line and place 186,000 mps in the middle, would put it to the minus side of the number line.

Whatever this pure energy is, ***IT*** will not slow down to the point at which ***IT*** will stop.

Things are not really going faster. Their speed should also be less than 186,000 mps. ***IT*** exists at 186,000 miles per second. At this moment, that is the fastest speed we know of that exists. Any slower speed is a speed lower than where time stops. The faster we go, the closer we get to the point where we say time stops. We are traveling as existence in a zone where the speed is less than where we say time stops.

❁~~~~~~❁

We can understand what IT was before IT reshaped ITSELF.

⌘~~~~~~⌘

Imagine that you are traveling on a jet airplane, and as you are looking outside the window it is like watching a movie. The slower the jet moves the more of the movie you will be able to see. The faster the jet moves the less you will be able to see. At the speed of light you would see nothing through the entire movie. It is because we are on this planet, which has a particular speed that we are allowed to see things. It is this slower speed, slower than where time stops, that permits us to understand the inner workings of ***IT***. We are watching this movie called "Life on Earth" at a speed that makes it possible to understand what this pure energy is. If the speed were slower we could not understand ***IT***, if it were faster we would not be able to observe ***IT***.

Time and memory

When we have experienced what we call an event that happened in the past, and we say that we can remember that event as if it just happened, it is because of this: What happened as a past event actually happened as this same moment that you are now living in. This is why you feel that you can remember the event as if it just happened. In reality, it happened as this same moment, but many of Earth's rotations ago. It happened in this same present moment but you cannot see it physically the way it happened then, because the event has changed as ***ITS*** weight.

In other words, the reason you cannot truly remember the event that took place many rotations ago as if it happened just now is because that moment and this moment are the same moment, but since that event changed, all you have now is the memory of it.

❀~~~~~~~~~~~~~~~~~~~~~ ❀~~~~~~~~~~~~~~~~~~~~~ ❀

IT does not reshape so that we can attach our human mechanical time system to IT **

⌘~~~~~~~~~~~~~~~~~~~~~~~~~~~~~~~~~~~~~~~⌘

Memories are very important in our human existence. When you went to school to learn, you had to use our human ability of memory, for memory starts when we are born, and ***IT*** incorporated this memory system long before we came into existence. In humans and other animals there is a racial memory which functions as instinct, which makes it possible for species to survive. During our first few years (rotations) we did not need to use our memory system to survive, for our parents protected us.

Our memory system is put to the test when we leave our protected environment. We leave to learn from others, such as teachers, so that we can survive after our parents are not around to help us. We strive to obtain diplomas so that we will be able to survive even better. Without them our chances of survival are weaker and more difficult.

We have to work hard in order to survive, so that we can take part in our second programming, which is reproduction. When this programming takes place we are forced to protect our children's survival, the same way our parents worked in order to guarantee ours.

On this note of life being harder without an institutional diploma, there are people that manage to survive this hardship so well that they develop street smarts, or get a "university of the streets" diploma, which comes only as an invisible diploma.

Learning is a necessity that ***IT*** incorporated into us so that we would have to work (transfer energy) in order to survive. In our transferring of energy ***IT*** searches for other existing possibilities.

Necessity is a very powerful force that ***IT*** placed in us to assure that we have to work, for in work we produce as ***IT***.

Getting back to memory and time: It is not really of great importance whether we realize that everything has been happening as a living existing moment where ***IT*** exists as a place, or if we continue to seeing things as if these events having a past and future attached to them, because this will not change the way ***IT*** operates. For ***IT*** only changes in this moment (a place) of existence, and is not affected by the human mechanical time system. So think about this: You have always, from the moment you were born, had this gift known as life, existed as the same moment called life, for you cannot exist anywhere else. What we see as different hours, days, months or years is only because of the way we were trained to see time. This training on time had to be really hammered into us so that the Industrial Revolution could take us to this existing point in our human development.

❁~~~~~~❁

** ***If we ask our mind if it has ever left this living existing moment, the answer would have to be no, because biologically and psychologically we exist from the moment that we are born, as being alive, and this is same moment that the mind has always existed in. Our minds cannot exist outside this living existing moment that IT exists as.***

⌘~~~~~~~~~~~~~~~⌘~~~~~~~~~~~~~~~~~~~~⌘~~~~~~~~~~~~~~~~⌘

And here is where our short human memory system does not help, due to our short stay on this planet. Many rotations ago we had to start living with a timepiece so that we could punch a clock (since most people did not have the money for clocks, many would be located around the cities and towns for public use) so that we could be where we needed to be in order to survive better.

Now we have to use the memory that is recorded on paper as history to understand that there was a moment when we did not have to have a timepiece in order to exist.

But no matter what you think, whether or not you agree with what I am saying, one thing is for sure, and that is that we can only be in this existing moment that you are reading this information, and if

ever you come back to re-reading this you will still be in the same living existing moment.

And if you think this is not so, try and see if you can remove yourself from this existing moment, you will find it impossible. The closest we come to detaching ourselves from this existing moment can be a very dangerous place, such as Alzheimer's disease and other diseases disconnect us from the reality of this existing moment.

When we say "I remember that event as if it just happened," the statement is true, but only as a memory.

The train station

There is a past, present, and future in our existing moment. Here is how it works: Imagine that you are traveling on a train or a bus. As you travel from one station to the next, you are experiencing past and future. The station you leave behind is the past. The station that you are approaching is the future. Think about this. The station that you left behind (in the past) is still there as you think about it. The next time that you pass the station from the past, it will be the present. It should be the past, but it is not because you are still in the same existing moment. It is the present because the planet has continued to rotate. What has actually happened is that you and the train have transferred energy in some form as the planet was rotating.

❁~~~~~~~~~~~~~~~~~~~~~~❁ ❁~~~~~~~~~~~~~~~~~~~~~❁

**** ***All changes are manifestations of ITS reshaping.*** ****

⌘~~~~~~⌘

Once we die, or the train is demolished, it will not be the past. It all remains as pure energy, reshaped into something else in omnipresent.

Outer space

We perceive outer space as starting as soon as we leave our planet Earth, and the first thing we notice when leaving the planet is that it is spinning. Our planet makes a full rotation in what we understand as 24 hours. As we look out into the Universe, we

know that everything that exists in space, including all planets, stars, meteors, etc., exists in an area that we refer to as being 95% empty space, which exists as a timeless area. This area is very similar to the atom, which is also 95% empty space. There is a good reason for this when we talk about some of the events that take place in this area.

First, we know that our human mechanical time system does not exist within the atom, so what happens inside the atom happens as events. All that takes place in our life and Universe within this omnipresence are also events that have been taking place since we came into existence, and will continue until and after our death. If you look closely, you will see that everything that happens in your life happens as events, which goes back to when we were in our primitive development, long before we established our human mechanical time system.

Let us say a star has exploded, and that it happened many light years ago. Accordingly, we believe what we are seeing is in the past; the light we are receiving is the light of the exploding star finally reaching us. I disagree with this theory. Recall the meaning of Omnipresent. According to the dictionary, omnipresent is that which is present in all places simultaneously. Omnipresent means that everything is within the existing moment. Everything that exists is one.

❁~~~~~~❁

****A vacation is when we try to change our existing moment as the routine we have to perform *******

⌘~~~~~~~~~~~~~~~~~~~~~~~~~~~~~~⌘

Follow me into this Universe that exists as omnipresent. In understanding omnipresent, God as pure energy, is the same as the whole Universe. Everything is happening in the same moment as a place; everything is happening in the existing moment as omnipresent. So, saying that to see the light of an exploded star is to see an event that happened previously runs counter to our definition of omnipresent.

The starlight we see is actually light from a star that does not exist as a star anymore. It exists as light energy being transferred through the process of transmutation. In other words, it is no longer a star; it is energy that is being transferred in distance as light. The starlight we see does not exist where the star used to be because it is no longer there. What we see is the star reshaped, after exploding, into light.

The star, as light, is now closer to us (in terms of distance). This light is taking place in the same moment that we are seeing it.

❁~~~~~~~~~~~~~~~~~~~~~~❁❁~~~~~~~~~~~~~~~~~~~~~~❁

****Pure Energy has no need for time, IT has no beginning or end***

⌘~~~~~~⌘

When you are physically in the presence of light you must remember that at the speed of light, time stops. When time stops, you are in this immediate moment, and this is what is called omnipresent (where only the existing moment exists). In the omnipresent there is no past or future, only the existing moment, which is where God has always existed as pure energy.

Try to see this whole Universe as one total package operating as omnipresent and as pure energy (God). You will see and understand why all ***IT*** has been doing is reshaping ***ITSELF*** in ***ITS*** existing moment. We take the whole Universe as one that exists as omnipresent. Travel from one part of this Universe to another will always be happening as the existing moment because of the omnipresent, which does not permit a past or a future.

When we travel from one place in the Universe to another we are covering distance in exchange for energy as fuel. Because everything is created in ***ITS*** own image, through the constant transference of energy, you will get change.

❁~~~~~~~~~~~~~~~~~~~~~~❁❁~~~~~~~~~~~~~~~~~~~~~~❁

*** ***Change is God as pure energy, as IT evolves**** ***

⌘~~~~~~⌘

To travel to another planet, a transfer of energy is required in exchange for distance; however, this will take place in the same existing moment. The time required to cover a distance will

depend upon how we have reshaped our ability to create speed in terms of technology at that moment. Even still, no matter where we travel, we will always be moving from one point of this God to another because we cannot exist outside this pure energy.

❁~~~~~~~~~~~~~~~~~~ ❁❁~~~~~~~~~~~~~~~~~~ ❁

Pure Energy has no need for time. IT has no beginning and no end

⌘~~~~~~⌘

A Moment

There is a saying that goes, "Do not leave for tomorrow what can be done today." There are several reasons for this saying. If something is done at the moment that it should be done, ***IT*** will continue reshaping at that moment. In leaving something until tomorrow, ***IT*** will have reshaped from the moment that you should have done it. The possibility of you doing it will not be the same.

Leave nothing until a so-called tomorrow, which really doesn't exist and that has never really existed. This is the same moment. In reshaping as humans, ***IT*** can exist in the surroundings that you now find yourself.

We should remember that the human mind gave a meaning to the word time. We even base its meaning on money when we say "time is money". And we've thought this way as since the beginning of the Industrial Revolution: The more you work in terms of time, the more you receive as money. This thought system is still in place and reinforcing our human concept of time.

A moment in time is the smallest fraction of that which we relate to as the mechanical time system. Sometimes we call it a second. Because this is the smallest fraction of our mechanical clock, we see and feel that we exist and are alive in this moment.

Before there was time, the moment existed as a place. This is the moment that gave us time. The existence of time is the same as the existence of the original dense matter at the moment of the Big Bang.

The only reason time exists is because the human mind says it exists. Before the human mind began, time did not exist the way we understand it. Before our solar system came to be, the time system we use did not apply. The only moment that has always been is the existing moment.

A moment is the same for every one of us. Your moment is the same as my moment. As I sit here at this computer, at this moment, the only things that I can do are the things that are in front of me at this moment.
I can continue to type or I can stop and look around to see what else there is that can be done within this moment.

❁~~~~~~~~~~~~~~~~~~~~❁❁~~~~~~~~~~~~~~~~~~~~❁

*** ***The existing moment is universal as omnipresent*** ***

⌘~~~~~~~~~~~~~~~~~~~~~~~~~~~~~~~~~~~~⌘

When we discuss the time required to get from here to the nearest galaxy, we are actually discussing how much fuel or energy it will take to get there in the fastest possible way. The amount of energy used will determine what we call human time, yet the traveling will occur in the same omnipresent moment.
Let us say that it is Monday and that on Wednesday, I will be taking an airplane from the airport. I know that this planet will have to make a rotation so that it could be Tuesday and another for it to be Wednesday. And this is when it is necessary that I be attentive to the mechanical clock, so that, like the airplane pilot, I will be in the same place as the same moment, so that I don't miss my flight!

❁~~~~~~~~~~~~~~~~~~~~❁❁~~~~~~~~~~~~~~~~~~~~❁

******All I can do is what is available to me as this moment*****

⌘~~~~~~~~~~~~~~~~~~~~~~~~~~~~~~~~~~~~⌘

Within the duality, I find that I want to do this and that. All of the things that I want to do are within this moment. Otherwise, my mind will fantasize. The more the mind fantasizes the more professional help the mind is going to need because the mind was made to exist in this moment.

Have you ever heard of a time machine? This is really an object put together by the human mind to take you away from the here

and now.

The other side of the duality is for me to do that which only I am supposed to do at this existing moment without asking why. I used to say that when it was related to the future, as in tomorrow, we had to wait for tomorrow to arrive. Now that I know there is no tomorrow and the only thing that exists is this moment, I have to rearrange the way I see things. I used to say, "Let me see what the future holds." Now that I live in the moment, I say, "I will have to see how this pure energy reshapes ***ITSELF*** as this existing moment." I will do that which I can do at this moment, no matter how many of Earth's rotations take place.

I can no longer say "We are wasting time." Time cannot be wasted because the reality is that time does not exist. Nor can I say, "I'll see you in a moment." I will actually see you in this same moment as this planet keeps turning.

Consider the expression, "time is running out." It's not. Something new is already in motion and this something will happen as this planet continues its rotation. There is no "until the end of time" but rather, how long ***IT*** will take to reshape into something new as ***ITSELF*** within the existing moment as a place.

It is not the human mind that has been operating the Universe; the human mind has just been limiting our ability to understand ***IT***. Humans, as an existence, transfer energy. We do this depending upon what is put in front of us in the form of possibilities that only we can execute in the moment.

If aliens exist they would be on a different mechanical time system. Their time system would depend upon what they use to keep track of movement. It is possible that they too could reach the understanding of only the existing moment. But they, as aliens, will have to exist as this same omnipresent moment.

Spin as time

As ***IT*** is a duality, the Universe is a model of the atom; and the atom is spinning near the speed of light. The effect of spinning (rotation) is very important. Remember, the Earth itself is spinning; it is also rotating around the Sun. The Sun is inside the Milky Way galaxy. The galaxy is also spinning faster than the Earth. The Universe, which is estimated to contain several hundred billion galaxies, is spinning even faster. And the outer edge of the Universe maybe spinning at the speed of light. This spinning is not always noticeable here on Earth. When we look at a rock that is millions of years old, it seems still, yet it is actually moving near the speed of light. Here is how: While the outer structure of the rock appears not to move, the internal structure of the rock is made of atoms. All atoms have electrons that are spinning near the speed of light.

Like rocks, our buildings and streets are considered straight and still, yet they too are spinning on our round planet. Additionally, a jet airplane does not travel in a straight line, for our planet is round and movement from one point to another will be in an arc. We see things as they happen over the curvature of our planet. Since the Universe is always spinning in the same direction, ***ITS*** evolution has always been in a forward direction. This allows for what we know as time. So far, ***IT*** has not changed gears and gone into reverse.

Time

We should start by seeing a moment not as human time, as in a second or a minute, but as a place in which ***IT*** reshapes, similar to what is happening within an atom, where events take place within a no-time system. All that happens within this place of omnipresence, from the beginning of the moment of our existence to the moment of our death, happens as events, not as time. This is to say, you exist wherever you may find yourself as a place and where you can do things as events. If these events take place in your home, your home exists as a place in your country; your country is on your planet; your planet is in your galaxy; your galaxy exists within the Universe as a place, and all of these events

are taking place as one moment we call omnipresent.

Yet the human mind forces its time system on ***IT*** and calls it reality. Try to visualize the following: Using the mechanical time system, the human mind has figured that this Universe is five billion or so years old. The mind counts time from where we now exist to where our time system would have first had the conditions necessary to operate. So it counts time by reversing the spin of our planet to see where it was just being formed. But remember, it is because of the forward spin that we know a day; it was the formation of our galaxy and the Moon that enabled the mind to conceive the idea of a month; and it is through the rotation of the Earth around the Sun that we know a year. If we reverse the spin of the Earth to where it began, we return as well to the beginning of our galaxy, where we would not have the conditions we now use in telling time, where we would not be able to use the words day, month, or year, let alone know their meanings.

Certain conditions have to exist in order for our time system to have a starting and stopping point. It is important to see that we have taken our time system to a point in which we force time on ***IT*** so that we can believe that our concept of time is absolute. We have given ***IT*** a time of birth, the same way we give everything a time and day of birth and death.

We should remember that when ***IT*** reshaped into this Universe, a system of time was not needed. There was no need for a star to be at work at a particular time--think daylight saving time or leap year--to satisfy our need for convenience. But will we ever accept that our human mechanical time system is only a convenience? Because in order to understand the Universe, (this pure energy that we know as GOD), we will need to open that door and accept that ***IT*** only exists in this place known as omnipresent as one moment of ***ITS*** existence

❀~~~~~~❀

****** It is not a question of whether or not you will live one more day. It is: Will you exist for one more of Earth's rotations? ******

⌘~~~~~~~~~~~~~~~~~~~~~~~~~~~~~~~~~~⌘

From a very early stage in our lives, we learn to think in terms of days; our present is "today." As kids we are told that "today we go to school" or "today we go on a picnic"; that today we will do this or that. From school days we shift to work days, and are told that "today we have this project " or "before we leave today, we must finish that project." People who work the night shift do not say they must accomplish their work "today," because for them, it is night. But just as we exist not in a day but rather as a moment, so is our work a transferring of energy as an existing moment.

While this is true still *"today"*, it is most clearly exemplified during our primitive development. Before our minds needed the convenience of the human mechanical time system we did not have to be at any particular place at a particular moment. We were governed by the sunlight that allowed us to see what we were doing and what was in front of us--when it was safer to hunt our food. Sunlight allowed us to see and flee from animals that would kill us if we did not see them first.

Sunlight, as a day, was a safer place to exist. So we began the day with sunlight and ended it with a resting stage that we know as sleep. During sleep, our minds reflected upon the events of daylight and seemingly conceived that these events happened before that night.

Because we have trouble accepting our mind at rest, we see ourselves as having worked in daylight, grown tired and rested, then continued again with the same light that has always been there. Additionally, we see events as having happened yesterday because we see our Sun as coming up and going down, even when we know this is an illusion resulting from the rotation of our planet, which is a result of ***IT*** putting spin and speed in everything that exists.

So our minds continue to see life's events as days and nights, which to the brain are different times due to their opposite natures of lightness and darkness. They are so different, in fact, that most of us use darkness to rest and daylight for work. But as we sleep

we are still transferring energy, only unlike during daylight, when we transfer energy to the outside environment, at night we transfer energy within our body. We do this in order to survive. We use darkness to rest, as did our primitive mind, but it did not yet label this Universe as a place that was governed by time but rather as a place where we transferred energy.

In our primitive stage, our minds did not attach time to the activities done as a day, week, or month. The beginning point of light upon waking happened not as a new day but as a moment where we had to work, that is, to transfer energy in order to survive. And we existed just the way we exist in this moment, as a place, not as time, where we are permitted to transfer energy as ***IT*** reshapes.

We are in the habit of saying that things started so many hours or years ago, or that something will happen in so many days or months from now, and they might indeed happen, but not because we are traveling along a timeline, but because both our planet and our galaxy are rotating. Our sense today of events as time and that there is a yesterday and a tomorrow results from our awareness of Earth's spin. Because ***IT*** put the world in motion, we see what is outside of us as moving, and our mind makes sense of this movement by describing it as backward and forward; thus giving us yesterday and tomorrow. The spin, however, is not based on time. It is based on an action that ***IT*** produced in order to reshape.

I have been insisting on this subject of time not because of time itself but because by removing time you will see ***IT*** more clearly. You should remember that ***IT*** does not exist as our mechanical time system or as a yesterday or a tomorrow. We have accustomed ourselves to believe we exist as a day, whether from our moment of sleep or because we wake and walk in the presence of light. You are not really living in a day; you exist in a living moment. ***IT*** is in this place called omnipresent, this very dynamic moment, in which we exist.

Time is based on speed

Whether time is slower or faster makes no difference. The common factor is that in all cases you will have lived in the same moment. This same moment as a place is the only thing that exists no matter where you are in the Universe as omnipresent.

Imagine that our planet always had sunlight and no darkness. If we had more than one sun, there could be constant sunlight. If our planet did not have the Sun to go around in its 365-day rotation, our mechanical clock system would be obsolete.

This would make strange things take place. By not having darkness, we would lose the meaning of one day. By not being able to measure a week in seven days, we could no longer measure months and years. We would have no way of determining our age in years. Our mechanical clocks would not have the same meaning of time. Our bodies would still be born and develop; they would go through the different stages of life until old age. Our only reality would be that we exist in this same moment.

If we had continuous sunlight, we would sleep when our bodies were tired. If we had continuous darkness the same would apply. We would awaken when our bodies were rested. We would lose our concept of time (Benedictine time) because there would be no measure to the beginning or ending of a day.

We are alive and were created in ***ITS*** own image, which is to be here in this exact moment. When our minds drift from the current situation or we lose touch with reality we are still in the same moment, but our minds choose to go elsewhere. In this situation, professionals may be necessary to bring us back to reality.

You may be familiar with some of the adages that relate to this notion of reality being in the present moment: "Keep your feet on the ground," "To live life fully, you have to enjoy the moment," "Live the moment; that is where everything happens," and "There's no time like the present."
The more you live your life in the past, the more likely you will get

tripped up in the existing moment. If you were to be extended in mid-air above New York City and stayed there for four hours, when you came down you would be in California because that is the rate at which Earth spins.

❁~~~~~~~~~~~~~~~~~~~~~❁❁~~~~~~~~~~~~~~~~~~~~~❁

****** Reality is the immediate moment, and this is where we were created to be ******

⌘~~~~~~~~~~~~~~⌘~~~~~~~~~~~~~~⌘~~~~~~~~~~~~~~⌘

Universal time

Let's travel into outer space. But first, let's understand that our mechanical clocks are not important when we leave this planet. Universal time is also the existing moment.

Let us say that you are born as a twin. One of you takes a trip into outer space while the other remains on Earth. The twin traveling in outer space is moving close to the speed of light. Let us say that the twin in space has traveled deep into outer space for the equivalent of 15 years on Earth. Meanwhile, the twin on Earth consumed a certain amount of air, water, and food in order to exist. The twin on Earth has been here for 5,475 rotations of the Earth around the Sun. Using our mechanical clock system to keep track of days, weeks, months, and years, this twin believes it is 15 years old. The twin in space did not take her mechanical clock. This twin did not experience the rotations of the Earth. This twin will not be able to relate to her age in human time.

The twins have things in common however.

They both took their first breath of air at the same omnipresent moment. They will experience more or less the same biological changes until they die. The twin that went to space will also consume air, water, and food in order to exist while she is traveling near the speed of light. This need for sustenance is what shows the biological similarities between the twins.

Time, for these hypothetical twins, is the same existing moment. They are both living in the same moment in time as omnipresent. The only thing that exists is the moment as universal

time. The twin that was traveling near the speed of light only gained distance. The moment for the twin in space is the same moment for the twin on Earth.

❁~~~~~~❁

****** The only time in the whole Universe is the existing omnipresent moment******

⌘~~~~~~~~~~~~~~⌘~~~~~~~~~~~~~~⌘~~~~~~~~~~~~~~⌘

If we see omnipresent as two hands, one hand is time and the other is a word called omnipresent. Let me give you an example of omnipresence.

First, however, return to the beginning of this book, and look again at the image of the young and old lady. This will help you to understand what I am about to say, for as you look at the picture, there are two extremely different views to perceive, yet what is most important about this is that they are both one; they are 100% omnipresent.

So now imagine that I have in my left hand that which we have been calling "time," which has functioned extremely well and conveniently as a system which we are all familiar, having been indoctrinated into it from the moment we were born.

And in my other hand we have a word called omnipresent, which we know exists at least in the dictionary, so meaning has already been given to "omnipresent," and as far as I can see, this is only a word that may or may not from here on have a more truthful significance as it relates to our human existence. Yet, if we focus more on this word and its meaning, certainly more about it will be uncovered regarding its relationship to ***IT*** as omnipresent, for this word enables us to understand ***IT*** as the pure energy that has always existed in this place called omnipresent.

Here is one more way to picture what I am talking about: If GOD is in all places at the same moment, this means that in my moment of existing, ***IT*** is here also. And if GOD is all knowing, then ***IT*** is also inside of me as this existing moment, firstly, as omnipresent,

and secondly, since ***IT***, to use the phrase, "created everything", ***IT*** is also you and I, and everything that may exist from ***ITS*** creation.

Finally, if you have or find more information on this word omnipresent that you would like to share with the rest of us, please send it to me at the contact information provided at the end of the book.

Universal law

There is a scientific law that there cannot be two things in exactly the same place at exactly the same moment. I want to show you how this relates to **IT** (GOD, pure energy) and the meaning of omnipresence.

Let us start with the meaning of the word omnipresence: ***IT*** is in exactly all places at (and existing as) exactly the same moment.

Now let us look at the scientific meaning of pure energy: it is the source of the existence of everything, from the minutest particle of the smallest atom to the vastness of the total Universe.

❀~~~~~~❀

*******Antiques are objects that have resisted being taken away from this existing omnipresent moment*******

⌘~~~~~~~~~~~~~~⌘~~~~~~~~~~~~~~⌘~~~~~~~~~~~~~~⌘

So, everything, you and I, are made of pure energy, which is made of matter and nothingness.

For this reason, the law that states that two things cannot be in the same place at the same time can also be restated as what we just finished talking about and the inevitable conclusion is that what we have is ***IT*** overseeing this Universe in being omnipresent and that everything is ***IT*** in all places at the same moment.

This may be hard for the human mind to see and accept. The problem began when we started using our mind to see and understand things as if they were completely outside of us. This is

a normal way of thinking because we actually see things as being out there.

Personally, it was when I fully understood the meaning of what we have been calling pure energy that I focused on what it meant that everything is omnipresent. I asked myself this question: "What if omnipresent is GOD?"

That is when the meaning of the word omnipresent became clear to me that everything is just ***IT*** reshaping as a place in the same eternal moment. Then I saw that we live in a three dimensional Universe and our mind provides the fourth dimension known as the human mechanical time system. ***IT*** exists as a place and we give it the name omnipresent.

Now I see everything as being one, pure energy, GOD, ***IT***. I exist in this omnipresence where ***IT*** has been reshaping what I see as a place. I exist somewhere as ***IT*** in this place that ***IT*** exists as. I have to exist somewhere. ***IT*** is huge. To deal with this immenseness we have come up with the concept of distance. Because of this we can understand ***IT*** better as far as size goes. We can understand and accept, for instance, that we are less than a grain of sand in relation to the Universe.

And speaking of sand, nano-technology could not have existed if not for it. There are different types of sand with different components, but quartz sand contains 46% silica, from whence come our silicon chips; not to mention Silicon Valley. And from the immenseness of the grain of sand we have gone into nano-technology, which remains ***IT***. We see and understand more as we delve deeper into the atom and keep "nanoing" further in, but it all remains ***IT*** in omnipresence.

But getting back to ***IT*** as a place where it existed long before the Big Bang occurred; scientifically, we have accepted that Time did not exist. Well, ***IT*** kept reshaping and doing fine without our mechanical time system. And more than that, ***IT*** existed as pure energy, so that we could come into being.

Since two things cannot exist in the same place at the same moment, I exist as ***IT***, pure energy and omnipresence, so do you.

I have to add that I cannot be grateful to anyone other than ***IT***. ***IT*** reshaped into my parents so that I could exist and learn more about ***IT*** with the information we now have. A hundred years ago (36,525 Earth rotations) the information I now share with you did not exist for the human mind to understand ***IT*** better, as to what ***IT*** is and how it operates.

All of the above is why I have accepted that this is a very gifted moment to exist in, and truly the only moment that ever has existed.

I can now look at ***IT*** as it was and as ***IT*** is, and as ***IT*** may become in all of ***ITS*** possibilities.

This omnipresent show is all about ***IT***, long before I came into existence and long after I leave, before and after everything.

Where time does not exist

It is an accepted scientific fact that our mechanical time system does not apply within the atom, which is the area governed by quantum physics. This is so because we need matter for time to exist, and what exists within the atom is not made of matter. Remember that what is happening as events inside the atom is also happening as that place in which ***IT*** exists as omnipresent; for whatever may exist inside the atom is also ***IT***. We should also remember that we, and every thing that is made up of matter, are made from this material that is timeless and free of our mechanical time system in its interior.

❁~~~~~~~~~~~~~~~~~~~~~~~~❁~~~~~~~~~~~~~~~~~~~~~~~~❁

We are made from material that has as its interior a no-time zone.

⌘~~~~~~~~~~~~~~⌘~~~~~~~~~~~~~~⌘~~~~~~~~~~~~~~⌘

Time as change

ITS reshaping creates the effect of change, which gives us the feeling of time, but time exists only as a moment. What we know as the future is only the immediate moment reshaping ***ITSELF***

into something new that will still be in the same place as the existing moment. We can experience the past in the existing moment when we hear and see the daily news, for these are events that have already taken place, but to accept the existence of the past violates the law of the omnipresent. It would mean that God would have used energy from the existing moment to leave behind as the past.

❁~~~~~~~~~~~~~~~~~~~~~❁~~~~~~~~~~~~~~~~~~~~~❁

The leaving behind of energy violates the meaning of omnipresent.

⌘~~~~~~~~~~~~~⌘~~~~~~~~~~~~~⌘~~~~~~~~~~~~~⌘

Our programming as time

We have been programmed since our earliest memory to think of things as one day to the next. When we wake up it is a new day. Since our first birthday we have become accustomed to seeing things in this manner. We believe that we are constantly moving forward. What is known as the past is what was left behind. When the Sun goes down, that day is in the past. Everyone thinks this way; it is natural to understand things in a manner which is comfortable to us.

I recommend that you experiment with and learn to see things as omnipresent, knowing that GOD, as pure energy, does not need time. ***IT*** is doing everything in this place called omnipresent.

IT does things in a repetitive manner. Although things are repetitive, each repetition is different because something has been reshaped, but it is all occurring as the same moment. As you read this, it is the same moment that the Big Bang occurred. It is not the moment that is changing; it is everything else due to ***ITS*** reshaping ***ITS*** weight.

❁~~~~~~~~~~~~~~~~~~~~❁❁~~~~~~~~~~~~~~~~~~~~❁

*** ***What we know as time is actually movement*** ***

⌘~~~~~~~~~~~~~⌘~~~~~~~~~~~~~⌘~~~~~~~~~~~~~⌘

The illusion of time

Because of the technology we know as communications, we as ***IT***, can physically be in New York at this moment and call someone on the telephone and within seconds "be" half way across the

planet. Our voice, energy transferred at the speed of sound, can be across the planet or in space.

When we telephone someone on the other side of the planet and the time difference in the country of the person being called is later than the area you are in, the person on the other end of the phone has aged more for that day than you have. Are they living in the past? What if you are in New York and are calling California? The person you are speaking with has not aged as much as you have for that day. Are they living in the future?

There is a place on the Earth called the International Time Line that, once crossed, will bring you to yesterday. Are those across the line living in the past?

With all of these situations, when you are talking to these people, or crossing the timeline it is the same moment. ***IT*** has no need for time. No matter where you are at this moment, it is the same moment at the other end of the Universe. ***IT*** is everything that exists.

The outside

Everything outside of us is made of atoms. Atoms are in constant motion. In motion, change takes place. What is actually happening is that **IT** is reshaping in that existing moment. There is no past; there is no future; everything takes place at this moment. The human mind sees change on the outside.

IT is in all places at the same moment; ***IT*** is merely reshaping ***ITS*** weight as the existing moment. ***IT*** has been doing everything within this moment. Before the Big Bang, ***IT*** existed as very dense matter. During the Big Bang ***IT*** reshaped into the elements. This transmutation has enabled ***IT*** to reshape into everything we know, which exists as atoms, such as you and I.

⌘~~~~~~~~~~~~~~~~~~~~~⌘~~~~~~~~~~~~~~~~~~~~~⌘

*** ***The visual effect of transmutation is what we have come to know as time.*** ***

⌘~~~~~~~~~~~~~~~⌘~~~~~~~~~~~~~~⌘~~~~~~~~~~~~~~⌘

The here and now

A psychologist asks a thirty-years old man to undergo hypnosis in order to take him back to the time he was ten years old. This can only be done in thought because the thirty-year-old man is in this existing moment. The ten-year-old child will exist in the place called the existing moment through the man's mind.

The now and then as omnipresent

Imagine that your present stage in life is a place in which you exist. Imagine that you could un-reshape yourself to where you were at the age of ten (according to the mechanical clock). Let us say that would be 3,640 Earth rotations ago. After seeing yourself 3,640 rotations ago, go back another 3,640 rotations. Continue to go back as far as you can remember. After completing this, remember that all of the things that have taken you from that point of existence happened in this place called omnipresent.

From the point at the beginning of your existence you were made in ***ITS*** own image. What has taken place is that you have continuously been reshaping to where you are now. Before you existed as a living being, your parents existed in the same place called omnipresent, the same place that you now find yourself in now, the place you know as this moment. Continue thinking in the reverse mode and you become more in contact with this pure energy. By continuing to un-reshape yourself you will see that you are the result of your parents' reshaping and their parents' and their parents', etc.

If you continue this process, you will go back to the point where no parents, or anything else that had life, existed. You will then realize that before life existed, this planet was just beginning to make its first rotation, which also happened in the same place called omnipresent, which is the same place in which you exist now. When the planet was just beginning, as a result of GOD's reshaping, it was beginning in omnipresent.

❁~~~~~~~~~~~~~~~~~~~~~~~~~~~~~~~~~~~~❁

**** Omnipresent does not exist as a yesterday or as a tomorrow****

⌘~~~~~~~~~~~~~⌘~~~~~~~~~~~~~⌘~~~~~~~~~~~~~⌘

If you go back far enough in this un-reshaping you will eventually get to before the occurrence of the Big Bang in this place called omnipresent. When I think of this very dense mass, I can understand that the only matter that existed was ***IT***. ***IT*** is not solid matter, however. If ***IT*** were, ***IT*** would be more difficult or even impossible to reshape into something like humans. This matter has mass and energy. We are not familiar with this type of energy.

IT has the intelligence that generates the reshaping of ***ITSELF*** from that very dense matter into the positive and negative energies required to produce an explosion, as ***ITSELF***. This explosion reshaped into all of the elements that now exist within this place called omnipresent. Let us be grateful that ***IT*** did, otherwise you and I would not be here.

❁~~~~~~❁

****** Our feeling of a tomorrow is also due to seeing darkness turn into light because of the Earth's rotation.******

⌘~~~~~~~~~~~~~~⌘~~~~~~~~~~~~~~⌘~~~~~~~~~~~~~~⌘

Astronauts

One rotation of the Earth is considered one day. Yet, if astronauts make one rotation around this planet in 90 minutes, that is not considered one day. Astronauts have been known to make 16 rotations in one day!

This is where we draw the line on time.

Astronauts have been asked if they felt "closer" to GOD in space. This question indicates that we still cannot see that we, as omnipresent, are this pure energy that is GOD. At some point we will become aware that this GOD, or pure energy, does not run on the human concept of the 24-hour day. At some point we will realize that something is not a million light years away; it is at the distance that exists as ***IT***, in ***ITS*** size, not as time.

We will be able to break this cycle of the 24-hour day when we leave the effect of Earth's rotation. This will happen when astronauts travel into space and do not return to Earth.

Remember, when you say that you will be here another ten years, what you are saying is that you will get to be a part of another 3,652.5 rotations of this planet. It was many rotations ago that primitive humans did not measure time. Primitive humans had no need to establish that a year was based on one full rotation of our planet around the Sun. Primitive humans had no need to attach an age to themselves. ***IT*** will continuously reshape regardless of the human mind's need for time. What matters is not how old you are, but rather how much you have been permitted to reshape in trying to reach your maximum development.

❁~~~~~~~~~~~~~~~~~~~~~~~~~~❁❁~~~~~~~~~~~~~~~~~~~~~~~~~~❁

Our concept of time is not required for human survival.

⌘~~~~~~~~~~~~~~⌘~~~~~~~~~~~~~~⌘~~~~~~~~~~~~~~⌘

The existing moment is not a time; it is a place where all events are taking place as ***IT*** reshapes ***ITSELF*** within this Universe. It is not millions and millions of years; it is billions and billions of times that this planet has rotated. This rotation is much more a reality than the abstract numbers we use in telling time. We may break this barrier of the time concept when we see evolution as GOD, or this pure energy, as a reshaping process existing as a place called omnipresent, not as time.

Astronauts still live by our mechanical clock because they return from space to the effect of our planetary rotation. Eventually they have to land at a particular hour in a particular place.

Perhaps someday our astronauts will be permitted to take that step into ***IT*** as outer space with a one-way ticket. If humans were able to use this one-way ticket, we would most likely begin to leave behind our mechanical time system. As we travel into the Universe that exists as ***IT***, as omnipresent, we would come into contact with ***IT*** as speed and distance.

IT would be harder for a crew of astronauts to keep saying that they will be in the next galaxy in the next 50 or 100 years. At some point they will realize that they are not influenced by Earth's rotation. But as they travel through this pure energy as omnipresent, astronauts will always calculate distance. Let's say

that the astronauts leave this planet and continue to communicate with Earth. As they get farther and farther away, they will continue to receive messages from Earth. The astronauts must remember that they, like the people on Earth, exist in the same moment as omnipresent.

Remember that the person sending the message from Earth is still alive, as are the people on the spacecraft, because they exist in the same Universe.

Regardless of the delay in receiving the message, it is still within omnipresent. The astronauts gained distance in exchange for energy as fuel. They may live longer because they are in a colder environment (outer space).

Cryonics

Cryonics is used in freezing people who have died with the hope of reviving them somewhere in the future, which will be the same moment as in omnipresent.

Let us say that Chrissine Rios died on January 1, 2000. When she died she was 30 years, or 10,957 Earth rotations, old. In order to maintain Chrissine's body as it was when she departed, she would have to be frozen. Pure energy in the form of electricity would also be needed to support the freezer that Chrissine would be kept in. Chrissine has to be frozen because heat is what keeps life changing in what looks like a forward direction. Cold, the opposite or duality of heat, stops the process of disintegration or decay as change. I have often wondered why this part of ***ITSELF***, as the Universe, is freezing cold. It would seem that this temperature exists to slow down the progression of what is happening in this Universe.

But, back to frozen Chrissine: She was hoping that some day she would be able to return to continue that part of her biological body that was not fulfilled. When or if Chrissine would return to life, she would survive upon that which has already been established in her DNA. The technology of the future that would make this

possible would be the pure energy that ***IT*** reshaped into that moment that we would know as "the latest technology".

Let's say that the technology required to be able to revive Chrissine was developed in the mechanical time year 3000. If this pure energy as GOD permits this to take place, and I do believe ***IT*** could happen, this could be better for Chrissine, as she would see changes that would be very interesting. It would be a way of taking our human bodies into outer space so that humans could continue as they are today. The other side of the coin is that in the year 3000, the human body may not be needed. Advancements in genetic engineering suggest we are headed in that direction now, being able to remove many of our current imperfections.

❁~~~~~~~~~~~~~~~~~~~~❁❁~~~~~~~~~~~~~~~~~~~~❁

*** ***IT put us together; IT can take us apart.*** ***

⌘~~~~~~~~~~~~~~~~~~~~~~~~~~~~~~~~~~~~⌘

Back again to Chrissine: For her to sleep for 1,000 years she would have to be here for 365,250 Earth rotations. This is actually a small number when we consider that this planet has a death date of approximately trillions of rotations before it will be consumed by a solar collapse; that is if everything works out ok for us. For Chrissine to exist for these 365,250 rotations we would have to supply energy in order to stop her maximization program. This would be accomplished by putting her in the freezer. ***IT***, as Chrissine, would have to continually consume more energy because that is in ***ITS*** own image, which is the constant transferring of energy.

When Chrissine wakes up in the year 3000 and looks around she would see changes that she feels have occurred since her death because ***IT*** continued reshaping while Chrissine was frozen. Yet if we could ask Chrissine her age she might say 30 because for her the moment she stopped breathing is the same moment that she continued breathing again, and she did so in the same existing moment as a place that exists in this Universe called omnipresent. The only thing that happened was that the planet made 365,250 rotations around the Sun; the planet continued as it should have. Consequently, ***ITS*** reshaping of her surroundings

would cause Chrissine to feel that she was no longer in the year 2000 or what she knew as the year 2000.

One thousand years after being frozen, Chrissine exists the same biological way she existed before she died, yet she is not one thousand years older. The moment that Chrissine died is the same moment that existed then, exists now, and would exist 365,250 rotations from now. Because our planet continues to spin, however, we apply time to the period of Chrissine's freezing. By applying time we are considering factors independent of omnipresence, which lead us to think of Chrissine's death as an occurrence of the past. But in the year 3000, ***IT*** is still the same place as omnipresent that ***IT*** was before Chrissine died. In order to reshape, ***IT*** does not need or use our mechanical time system.

Being free of time

I was trained to believe that time truly existed. I see this differently now that I know where, how, and why our mechanical time system was created and implemented. I am grateful to ***IT*** for giving me freedom of not having a boss or having to do any particular thing at any particular time. I am grateful to ***IT*** for freeing me of this cycle of the mechanical time system that I was born into. Now I am free from what I had known as time. I must thank ***IT***, God, as this pure energy for liberating me from the concept of time.

I now live with the use of this gift called my human body in this place that ***IT*** reshaped into as the Universe. This is where ***IT*** permitted Earth to exist so that I could live in omnipresent.

I live in a place where there is no yesterday or tomorrow. I live where there is no Monday through Sunday. I am not getting older; I am in the process of reaching my maximum development.

I go through the two basic programming functions that all humans go through. As mentioned before, the two functions are survival and reproduction. Now that I have fulfilled the second program in participating in the formation of two children, danger arises.

Fulfillment of the second program would seem to indicate that I have reached my maximum development; that is when we should be ready to die. And, many people feel that after they are 35 years old everything starts going downhill.

I had a near death encounter in a car accident. Because of this I started doing something that not everybody does after 35; I started taking care of my body. I began to exercise to relieve the pain caused by my accident. Exercise makes me feel much better; had I known how good I could feel I would have started exercising much sooner. I believe that by exercising I am slowing down the process of reaching my maximum. I am transferring energy within my own body. This transference of energy allows me to aim for a new maximum. This new maximum can only be obtained by me.

⌘~~~~~~~~~~~~~~~~⌘~~~~~~~~~~~~~~~~~⌘

** ***IT does not have to exist as time.*** **

❁~~~~~~~~~~ ❁ ~~~~~~~~~~❁

I want to stay on this subject to show that the mind also has to do with the physical. Since I have started putting my thoughts on paper, I have become aware that by using my mind I have forced it to continue this search for a new maximum. This project started with me writing down trivial facts. I then started the Big Bang section. While writing it I realized that I needed to go back to before the Big Bang for it to make sense.

That brought me closer to the understanding of omnipresent. In continuing the Big Bang section I came to understand the existing moment.

We should never allow ourselves to stop growing physically or mentally. As long as we continue to grow, we delay the process of reaching our maximization. Growth, mentally and physically, is the transference of energy. The transference of energy is, in essence, the reshaping of the existing moment.

We feel and see change. It is understandable why we say it is a new day or a new year. These things occur because ***IT*** reshaped and made this change possible. This began before the Big

Bang. When we say that this Universe is billions of years old, we are using the vocabulary we developed because of our existence and the existence of our planet and galaxy. Our concept of time will end when our galaxy destroys itself in order to reshape into something different. We know that this destruction is inevitable and that it will be the end of the time system our mind has created.

I used to say "Thank you God for one more day." Now I say "Thank you for permitting me to continue being here in this place that exists as the here and now." As mentioned earlier, it is not that time is running out; it is that since something is already in motion, something will happen as this planet continues its rotation, and it's already made trillions of rotations since it came into being. We can thank God that our planet is still in such excellent condition. It looks good and as new and fresh as ever. If I were a planetary doctor, I would say that ***IT***, as this planet, is ticking well and looking as though it could make a few more trillion rotations.

We need to have more patience; everything is happening as a singular moment and place where ***IT*** can occur. If something is going to happen in your life, it will happen; all you have to do is keep the road clear so that it will happen in the smoothest way possible.

IT as 186,000 mps

Let us take into consideration our use of the human mechanical time system to state that time stops. We base the theory of speed and time on our understanding that there is a past and a future, which as of this moment no one has been able to find to be true. This is why I find that the scientific belief that at the speed of light, 186,000 miles per second (186,000 mps), time stops needs correcting.

We have used the speed of 186,000 mph as a way to measure the Universe as it relates to distance, which is sensible with respect to speed and distance. But it is not sensible with respect to the Universe being ***IT***.

Regarding time, remember that we exist in a place known as omnipresent, where everything is happening as a place that exists as ***IT***, where ***IT*** can reshape as events. Therefore, I no longer believe that time stops at 186,000 mps, because in this place of omnipresent, time does not exist, nor did it exist before humans arrived, or before the Big Bang.

I understand that in order for my mind to reach your mind with words I have to use a vocabulary that we have all been accustomed to hearing, but I found that the words we use to relate to our human mechanical time system do not fit into this place known as omnipresent, and it takes an adjustment of thinking when it comes to time. In writing this book, I never expected that I would have to change the way I understood time, for I too was educated to think that time stops at 186,000 mps.

But now I understand that as omnipresent, "time" never existed, and that our human mechanical time system was put together by human minds as a human convenience – and it is an excellent system – that works effectively within this omnipresence.

Lucky for us, as a civilization that began without the need for a mechanical time system in order to function, we still exist in this omnipresence and as this moment. If we as a civilization disappear totally, our time system would not change the way ***IT*** is.

Accordingly, speed and spin have more importance and value than our human mechanical time system. In spinning slower than the speed of 186,000 mps, ***IT*** made our planet's existence possible, enabling us to see, talk, and think about ***IT***, even if all of this is just a human thought.

I suspect your mind, as mine did, finds this difficult to accept, so let me explain further. Let us begin with you. In order for you to read these lines, you need a human body that has a mind that can think. Your body is composed of matter, and all that is matter is composed of atoms, and all atoms are ***IT*** as a result of ***ITS*** reshaping into this Universe.

In order to think, you use atoms as energy, and for you to speak, again you use energy; speech is a vibration that produces a sound that can be heard as a word. In short, all of you is ***IT***, first as omnipresent and then as a name that uses atoms as ***IT*** to think, which is only a thought that is possible because ***IT*** exists in ***ITS*** reshaping as you.

When I became aware of this, all I could do was be grateful that ***IT*** exists and ***IT*** permitted me to exist, even as a thought that allows me to think, that I am Ric Ricardo who can think, see, feel, and enjoy this moment of ***ITS*** reshaping as life.

But let us look again at 186,000 mps as ***IT*** and its relationship to matter. Our scientific and human understanding is that at 186,000 miles per second matter cannot exist. This is so because matter, as atoms, has density and therefore weight, that will automatically slow down this speed of 186,000 mps, thus making this planet and our existence possible.

Because of my personal experience I know that the human mind can begin to understand that our mechanical time system does not exist within this Universe as omnipresent and that before our existence this mechanical time system could not be applied as ever existing. But the speed of 186,000 mps has always existed; both before and after the Big Bang.

Here again, to use language to say that at 186,000 mps is ***IT*** and time does not stop only serves as a way for a reader to understand what I am saying, by using words. Still, it is a starting point of understanding that the human mind, which has felt speed and spin, would think that since spin and speed do exist we must be moving in a forward direction and that there must then be something up ahead, and logically if there is something up ahead, then there must also be something left behind, leading the mind to reason that if something is behind there is also a past, and if there is a past, this would reinforce the concept of a future.

Additionally, I can see that in our primitive stage of thinking we did not know that this place of omnipresence existed. But now that we have more technical information on ***IT*** and why ***IT*** is pure energy, we can say that ***IT*** is as fast as 186,000 mps as omnipresent, and ***IT*** is also as slow as matter, for ***IT*** uses atoms to reshape into something that is made of matter.

If it were visually possible to see every thing that exists as ***IT*** within this Universe, we would only be able to see ***IT*** as the 5% that exists as matter, even if what we see is 95% hollow. Furthermore, to see this Universe we would have to see it at the slower speed which we know as the existing moment. So, remember that at 186,000 mps matter does not exist, but we need matter in order to attach our human mechanical time system to it.

A timeless place

The human practice called meditation is how we can be with ***IT*** as a timeless place that exists within us. A timeless peaceful place that lasts longer and longer, but there is a catch to it, and that is that we have to isolate ourselves from this material world to which we believe we have such a birthright.

⌘~~~~~~~~~~~~~~~~~~~~⌘~~~~~~~~~~~~~~~~~~~~~⌘

****Our day and night do not exist in outer space or for aliens*.***

⌘~~~~~~~~~~~~~~⌘~~~~~~~~~~~~~~⌘~~~~~~~~~~~~~~⌘

Why does time go fast and why do we say that?

What we are seeing in the past we are seeing it as if it happened this moment because we have seen a lot. We are experiencing the seeing of the many things we have done as the same living existing moment. We know that we are here to witness many events. We are aware that we had a beginning and that we will have an end. We compress our sense of time as we experience what we have seen.

And since we have experienced many events at, let's say, the age of 25, we know that a lot has happened, but since we try to fit these events into our mechanical time system, we again feel that time goes by fast, but the reality is that since our birth, and this is

something to think about, we have always been existing as what I call **a living existing moment.** I use this wording because with it I find it easier to understand the idea behind it.

The mind knows that in order for it to exist it has to be alive. Being alive makes possible our feelings and perceptions about time; how it goes by fast or slow. I say feelings and perceptions because that is all they are. The wonder of it all is that we have our minds; and for that I give thanks to ***IT***, as pure energy that reshaped so that we can be what we are, beings that think and feel.

So, we can see what we were before and we can see our process of reshaping into what we are.

I have been able to look back as far as to when I was two years old. As I look back and think about it I know that what I see are only events that happened. Everything is just a thought.

I know when those events took place on the basis of our mechanical time system, using my birth date as a reference point.

I am having these thoughts because I am trying to see that time in the reality that those events took place in this same moment of my present life.

We know that we have been here for many of Earth's rotations, and in taking in, or remembering as many events as possible at this moment, we get the feeling that time is going or has gone by fast.

But remember this, all these feelings about the passage of time can only happen in the moment that you are thinking about it, for everything that ever happened in your life has happened in this same moment in which you exist. It will also be the same moment when you die.

The ingredients for time to exist

Here are a few things that we should remember that are related to the existence of time:

For time to exist there has to be these ingredients:

1. Our planet Earth and its rotation existing at the speed that it is now spinning.
2. Our own existence, otherwise, who else is going to state that time exists?
3. The existence of speed, for it is the speed of things that gives our mind the illusion that something is moving. Then we can ask ourselves the question: How long has this movement been going on, and how fast is it going?
4. Matter, because without matter we or whatever is being timed won't exist.

This is why it is scientifically known that before the Big Bang time did not exist.

This is obvious when we look at the list above, and then look at what existed just before the Big Bang occurred:

1. Our planet did not exist.
2. We did not exist.
3. Nothing was moving. ***IT*** existed in a singular stationary position as very dense matter.
4. Matter, as atoms, was not yet formed.

So now you can see better why time is only a human invention and that ***IT*** does not exist as time, for ***IT*** only exists in this existing moment and not as a moment in time, but rather as a place, for ***IT*** is not like us, who feel that we have to leave something behind to confirm our existence. ***IT*** does not need to confirm to anyone that ***IT*** exists.

Life and death in the existing moment
You will become aware that you only exist in the moment you are in, and since you are aware that there is no tomorrow, or a time that can be called the future, you will come more in contact with life itself. This should also make you aware that death will also happen in this moment.

Our illusion is that we have a tomorrow, that we have more time, so death is further away. You will lose this illusion in exchange for living only in this existing moment. Everything happens now, in this moment in which we exist.

❀~~~~~~❀

Take care of your GIFT, known as this moment that you exist as ITS life; by taking care of your body and mind, for IT gave you the gift of a mind: how you use it will be your gift to yourself.

⌘~~~~~~~~~~~~~~~~~~~~~⌘~~~~~~~~~~~~~~~~~~~~~⌘

A timeless travel

Here is something to think about that is related to time and ***ITS*** nothingness. Let us say that you are traveling in, or better still, through, outer space. As you look out the ship's window you will see other celestial bodies that are moving. You are also moving as you cover distance in outer space. Here is the thing: For time to exist we know that we need matter, and since there are all those celestial bodies out there, as matter, we can see them moving, and so are we, because, like the ship and the celestial bodies, we are made of matter, and without matter, which is just ***ITS*** weight, we, like time, cannot exist.

I have been saying all of the above because, without ***ITS*** weight, we and everything in outer space cannot exist. This is a normal way to think, but like I have said before, we need to stop using our normal way of seeing and understanding things so that we can understand ***IT*** as ***ITS*** nothingness. In order for us to understand ***IT*** better we will need to focus on ***ITS*** 99.99% nothingness. This will also help you develop a stronger thinking mind, for you will be exercising your mind in a way that it was not being exercised before. You will not go crazy thinking of a nothingness, for all you have to remember is that this nothingness is there as ***IT***, and that ***IT*** is a very powerful force that has no competition, for ***IT*** is one of a kind that exists as a duality in ***ITS*** 99.99% cold, clear, invisible shell and ***ITS*** less than 1% weight.

Let me add to this nothingness what we already know as ***ITS*** opposite, which is ***ITS*** weight. I say this because if you look around at all the things that exist as ***ITS*** weight, you will see that

they are extraordinary: Beginning with our Sun to all the planets in our solar system, to the extraordinary human body that you occupy as ***ITS*** weight. So that we can use our very limited viewing power to see as much as we can of what ***IT*** exists in terms of ***ITS*** weight.

If you are like me and find that all of the above activities, as ***ITS*** less than 1% weight, are extraordinary, then let us see what this other 99.99% cold, clear, invisible shell that ***IT*** exists as is. Since this area of ***ITSELF*** is an invisible force we will have to use our imagination and equipment that can focus on ***ITS*** clear, cold, invisible speed.

Let us get back to our trip in outer space. Think about this: As you and your ship are traveling, you are in ***ITS*** 99.99% nothingness, and as nothing, there is no time system, or at least not our human mechanical time system that we can attach to something that exists as a constant nothingness. We need matter and rotation to be able to use our human mechanical time system.

Now that we know that we need matter in order to attach time to something (***ITS*** weight), we will do so from this place that exists timelessly, because, we will never be able to attach time to something that exists as nothing. But this is only something to think about, for it will be easier to digest it if you remember that we, like ***IT***, only exist as a moment, which is really just a place where ***IT*** exists as an invisible shell housing that is made of an invisible energy force that exists as a cold, clear nothingness that has expansion. We have, at least, begun to confirm what little distance we can observe in this cold, clear nothingness that ***IT*** exists as in outer space.

By the way, referring to *outer space* is actually incorrect terminology. We say *outer space* because we are leaving what we know as *home,* something we think of as being *inside.* Since we are leaving our home and going outside we use the term *outer space.* But now we know that what is out there is pure energy, or GOD, we will have to adjust our thinking and remember that we

exist within this pure energy as GOD as ***ITS*** invisible shell housing known as this Universe. We are not going into outer space; we are going to continue existing within ***ITS*** total shell that is made of this timeless nothingness. As we travel away from our planet we will, at some moment, see our 24-hour rotating planet disappear, and even our atomic clocks will not be able to reach us as we depart from our concept of time and matter. We will exist in ***ITS*** more than 99% timeless, clear, cold nothingness. To which I have to say, we should be grateful to ***IT***, for imagine if we left ***IT*** as ***IT*** exists, which will never really be possible because we would then not exist either.

❀~~~~~~~~~~~~~~~~~~~~~~~~❀❀~~~~~~~~~~~~~~~~~~~~~~~~❀

******The past and the future can exist only as a human thought that can only exist as a present moment.******

⌘~~~~~~~~~~~~~~⌘~~~~~~~~~~~~~~⌘~~~~~~~~~~~~~~⌘

Heat as time

This will sound funny, but there is a relationship between heat and time. To see this we should know a few things first. For instance, for the idea of time to apply to anything we need a couple of things: something has to be changing or in motion and for this to happen there has to be matter. We know that matter comes from the original pre-Big Bang dense matter. All matter comes accompanied by the original pure energy and heat that produced the Big Bang. We should recall that pure energy cannot be created or destroyed; so that the original energy is what we have today.

For us to say that something has time, we have to relate it to something material, which has atoms with spinning electrons and protons. We could say that heat gives us the ingredient with which we can apply time to something. Let us not forget, the opposite of heat is cold, which is a never changing constant. Heat gives off motion; we know that mass has heat, like in the center of our planet. Solar light is also heat, and we know that light travels at a particular speed, which means there is time and matter attached to it.

Why there is no future in a day

Here is something to think about when it relates to time as the future. When we talk to someone and we say to them, or relate to

them something that is going to happen as the future, we rarely ever say that what is going to happen as this future event will happen as moments away, or even hours away, we instead say that we will see each other a little later, but not as "the" future.

And we say this because, we know that if it is 9am and we will see some one at 11am, we would rather say that we will see that person later on, and even if we said that we will see that person, let's say at 11.45 pm, we would rather say that we will see them tonight, but again, not as a future event.

I have been giving you these examples so that you can see that for us to say something will happen as the future we need to at least stop our minds as this moment of existing and close down the mind as sleep, so that we can then wake up again and say that before we went to sleep is now in the past, and that for us to say something will happen as the future, we need to stop our minds from seeing what we have been seeing as the same existing moment, in order to call it the future.

You see, awakening from sleep is what brings us back to our living existing moment, because it is during sleep that some of us depart from existing in this living existing moment. When we awake, we again come into contact with existing as a moment that can only exist for us if we are alive. By the way, we have become so accustomed to it that we have forgotten this gift known as being alive.

Getting back to when we awake and that 11:45 pm appointment that we hope we can keep, and why we refer to it as later: This happens because we know, or our mind knows, that this 11:45 pm appointment will have to happen as the same living existing moment that exists as when the mind woke up. Our minds are not that stupid or dumb to get fooled into that 11:45 pm appointment as being in the future. The mind knows that that appointment has to happen in this moment that it (the mind) is still wake. The mind knows the dangers that exist if it leaves this living existing moment. This could be the reason why we have to be so tired

before going to sleep. If it was up to the mind it would continue staying awake, for the mind knows that if it closed down for a moment, as life, it stops existing.

Getting back to us following a living moment that exists when we awake, to that 11:45 pm appointment: We know that between the moment that leads us to that 11:45 pm appointment, many things will happen as ***IT*** reshapes, so that we (our minds) who exist only as an existing moment will be seeing many things happen that will eventually bring us (our minds) to that 11:45 pm appointment. To the mind this will all have to happen as it is still awake, not as time, because when we existed as primitives, or when our time system did not yet exist, this way of our mind existing would still have been true.

Our mind, when awake, knows that the things that will happen, will happen as the same living existing moment that it (the mind) now exists in, or exists as. Our minds have found it easier to say "later", since it (the mind) did not yet have the information that now exists about time being only a convenience, so that our minds know all the events that will lead it to the 11:45 pm appointment are going to happen as a moment that it (the mind) exists as. Our minds know that things outside of us are going to be changing due to people moving about, and because the lighting (our solar light) also changes, (due to Earth's rotation); so to the mind it feels safe to say that these changes are going to take place, and that there will be no danger in saying "later." That later appointment has to happen as the same moment that it (the mind) exists as, that is, as being awake in, so that when we now refer to something happening later, what we are really saying is that we know that between now and what we refer to as later we have to participate in ***ITS*** reshaping as it changes in ***ITS*** existing moment.

Time travel

There is a major problem here, and it is this: As much as the human mind has tried to remove us from the existing moment, and even in physics the mind feels that one could send subatomic particles into the past or future, ***IT*** only exists in all places as the

same existing moment. If ***IT*** had to satisfy the mind's past and future ***IT*** would have to fragment ***ITSELF*** in such a way as to use its existing moment, as energy, to send things back and forth to places that do not exist. This would also mean that ***IT*** would need psychiatric adjustment in order for it to continue existing in all places at the same moment and to still find time to go and be with the past and future.

But I have to say that the above situation does not exist, and that one thing that will make things clearer is to remember that scientifically we know that before the Big Bang time did not exist, and that **time is only a human invention, a mere human convenience, and presently our mechanical time system exists with an economical foundation attached to it.** And to make this clearer still, I can assure you that ***IT*** does not wear a timepiece, for ***IT*** does not need to leave the existing moment.

As for me, I have stopped trying to predict what ***IT*** will reshape into as this existing moment, into what others call the future, for I prefer to continue enjoying this existing moment which is where I can be as close to ***IT*** as ***IT*** exists.

⌘~~~~~~~~~~~~~~~~⌘⌘~~~~~~~~~~~~~~~~⌘

*** ***PREDICTING THE FUTURE MEANS THAT YOU WILL NOT HAVE A FUTURE, FOR YOU WOULD BE EXPOSITING A PLACE AND TIME WHERE YOU CANNOT EXIST WITHIN THE OMNIPRESENT*** ****

~~~~~~

IT too obeys ITS rules

Every thing that is happening throughout this Universe is happening as ***ITS*** existence, and everything is also happening to ***ITSELF***, as a singular moment, as everything that ***IT*** reshapes into. Think about this: Everything is reshaping as the same moment of ***ITS*** existence, and what is reshaping is only happening to ***ITS*** heated weight. Let me explain it this way: Since ***IT*** is **ONE**; it can only change, as ***ITSELF***, as a singular, continuous moment.
~~~~~~

Let us look at how ***IT*** now exists throughout the Universe. The only way ***IT*** can reshape ***ITSELF*** as the same moment in ***ITS*** existence is to exist as ***IT*** now exists; fragmented throughout ***ITS*** shell (Universe) where ***IT*** can change into so many different things. ***IT*** can only change ***ITSELF*** as the same moment in ***ITS*** existence, which is what ***IT*** is now doing as all the things that ***IT*** exists as that have ***ITS*** fragmented weight attached to ***IT***; planets, for instance, and everything that also has ***ITS*** fragmented weight on the planet (like us), and in everything on the Moon, and as in everything that is happening in the Sun, and all the things that I have not mentioned that exist within ***ITSELF*** that are made from ***ITS*** fragmented weight.

IT too obeys ***ITS*** own rules: Everything that is happening is happening as the same moment, or, to not to use the word *moment* meaning time, but is happening to ***ITSELF*** as it exists.

If we remove our mechanical time system from everything that is happening within this Universe (***ITS*** shell), we will see that everything has been happening to ***ITSELF*** as the same moment; which is really a place. Before and after the Big Bang everything has been happening as the same moment (or place) of ***ITS*** existence. If we recall before the Big Bang, when ***IT*** had all its heated weight in one place within ***ITSELF***, time did not exist, and it was after the Big Bang that ***IT*** threw out its weight as quantifiable fragments (atoms) throughout ***ITS*** invisible shell that ***IT*** reshaped into. We, after many rotations, came into existence and invented the clock, using our planet's rotation as a measuring device for time. This is just a human convenience.

IT obeys its own way of existing, in which ***IT*** can only change or reshape as the same moment (to use a word) of ***ITS*** existence. This will make more sense if we go back to understanding ***IT*** when ***IT*** had all its weight in one place, before the Big Bang, when ***IT*** existed as very dense matter. We must remember that ***IT*** is only changing as the same moment. When ***IT*** had all its weight in one place ***IT*** had fewer possibilities to change into, ***IT*** had only the substance (***ITS*** heated weight) that ***IT*** is made of.

When ***IT*** searched for other possibilities to reshape into, through that part of ***ITSELF*** that makes change possible in quantifiable portions (atoms) as ***ITS*** weight throughout ***ITS*** invisible shell that we call the Universe, ***IT*** all had to exist within ***ITSELF***, and ***IT*** all had to change at the same moment as ***ITSELF***, within ***ITSELF*** in ***ITS*** only existence.

Maybe you will see it better this way: As omnipresent, ***IT*** is in all places as the same moment, for ***IT is*** all the places that exist as ***ITS*** weight, as ***ITSELF***, and ***IT*** keeps reshaping ***ITS*** weight, that exists within ***ITS*** invisible shell, as the same moment; for nothing is really changing before or after ***ITS*** existence.

❀~~~~~~❀

******We are all born in the same omnipresent moment, but we arrive and leave on different Earth rotations ******

⌘~~~~~~~~~~~~~~~~~~⌘⌘~~~~~~~~~~~~~~~~~~⌘

What makes a time machine?

Here is one more way to understand why it is a difficult situation to put together a time machine that could take you out of your existing moment into, let's say, the past. Now, first, it is very important that you remember that time cannot exist in our mode of existence without what is called matter, so that we could see other things (objects that are made of matter) that also have to exist as matter, so that a situation could exist where we could apply time to us and the object in question.

❀~~~~~~~~~~~~~~~~~~ ~~~~~~~~~~~~~~~~~~~~❀

****** We cannot apply time to ITS cold clear invisible shell body, for it is made of a form of a nothingness ******

⌘~~~~~~⌘

Now, if we start by first dealing with the time machine that will take us back in time, before we even get into this time machine, we will notice that this machine is made of something called matter, and since we are now in a high tech society, this time machine will most likely have many computers inside to help us try to make this jump back in time, so that this time machine with everything inside of it, will have to have weight as the matter that it is made from.

After noticing this, let me show you why there will be problems with this journey into the past. Since we are now so technologically advanced, we can go farther back than the journeys that we have seen on TV, such as going back to 18th century. We are going to go back to just before the moment of the Big Bang, so that we can see why there are problems in our leaving our existing moment because we, like everything else that may exist that is made of matter, is made from ***ITS*** weight, and this is more important than the subject of time itself.

To continue: If there were a time machine that could return to the past, what has to happen is that as this time machine is going back in what we call time, this machine would have to be able to do something that we are not allowed to do, which is to un-reshape the things that now exist as matter!

❀~~~~~~❀

**** ***IT does not sleep for IT is not governed by time*** ****

⌘~~~~~~~~~~~~~~~~~~~~~~~~~~~~~~~⌘

As this machine is un-reshaping what ***IT*** reshaped from in order to go back in time, when this time machine (supposedly) arrives at the moment of the Big Bang, what will happen is that the machine and its traveler will not exist, and the reason for this is that everything that now exists came from ***ITS*** weight when ***IT*** had all ***ITS*** weight as one singular point that existed within ***ITSELF***, when we, like the time machine and like our planet, and even our galaxy had not yet been shaped from ***ITS*** weight. In the same way, if we tried to travel into the future we would have to wait until ***IT*** uses ***ITS*** weight that now exists to reshape into what we call the future.

So again I say to ***IT:*** Thank you for allowing us to see and understand the way you exist, as your constant, empty nothingness, where you have your ever changing weight, that you have as your constant weight that exists inside of you. And if you, the reader, are wondering why I said ***ITS*** constant weight, it is because ***IT*** does not gain or lose weight the way we do, as ***ITS*** total inside weight, for ***ITS*** outside nothingness does not have weight attached to this area, simply because ***ITS*** outside is not made of something that has to have weight.

Space-time

Here is something for us to think about, when we refer to space-time: Why not just see this space-time as ***IT*** ? Since our minds have been conditioned to see time through our mechanical time system then let us see ***IT*** as space-time. This can be accomplished by visualizing the fact that there is indeed something that exists as space, and this space does have distance, which is where we use our mechanical time system to measure how long something (***ITS*** weight) takes to get from one place to another.

You can try this, but you will need to use your imagination, and remember that imagination is when you look inside yourself to see something that is not there physically, or something that doesn't yet exist... This is one quality that engineers that build things have, for they have to picture in their minds, to use an example let's say a construction builder, who has to visualize on the basis of what is on a blueprint, what the final building will be like.

So using your imagination picture yourself standing outside the now existing Universe and looking into what now exists inside this Universe. If you do this what you would see the many kinds of celestial bodies that exist, that are made from ***ITS*** weight, that are inside this place that we refer to as the empty Universe. Now, what is important is that we, like ***IT***, are seeing everything that exists inside this Universe as just one moment in ***ITS*** existence, that we call omnipresence. Perhaps it will be easier for you to see it this way: When you look at everything that exists as ***ITS*** weight that exists inside of ***ITSELF***, everything inside exists as one frozen moment called omnipresence as what we understand as all the matter that exists within the Universe as ***ITS*** nothingness, and we would also have to include everything that can happen inside the tiniest single atom, for this atom also exists inside and outside as ***ITSELF*** as one frozen moment of ***ITS*** existence as omnipresence.

Unfortunately, one thing that we can never see is ***ITS*** high speeded nothingness (***MAXX-SPEED***) and this is because this speed cannot have anything attached to it, for then it would be part of what we call matter, Here, too, we can use our understanding of the phrase

that time stops at the speed of light. If this is so, then it would mean that since this ***MAXX -SPEED*** is faster then the speed of light, our mechanical time system would not function, the same way time cannot exist, if ***ITS*** weight were not fragmented, so we could process the distance that exists from one fragment of ***ITS*** weight to another as omnipresent. So remember that as you view this Universe, every thing that is moving is happening as omnipresence.

Have you ever considered that ***IT*** does not need time to measure the distance from one fragment of ***ITS*** own weight to another fragment of ***ITS*** own weight, because all ***ITS*** weight, exists within ***ITS MAXX- SPEEDED*** nothingness, (where time could not exist anyway) that exists as ***ITS*** body, and it was ***IT*** that put its fragmented weight into motion when ***IT*** reshaped into the Big Bang? And it is from this moment on that we can apply our mechanical time system: first, because we came to exist, and second, because ***ITS*** fragmented weight has distance, where we can measure the different distances that exist from fragment to fragment, that exist inside of ***ITS*** timeless body.

Our reference to time

When we say to someone that we will see them "some other time", what this really means is that they will have to see each other again in this same living existing moment, but so many rotations later. This also applies to when we say that we will see them "in a little while", because when we see that person we will be seeing them in the same moment that exists as both our lives or as the same living, existing moment. You will better understand this when you remember that we are alive because ***IT*** is alive, as omnipresent, and that ***IT*** does not need to exist as our mechanical time system nor does ***IT*** need our mechanical time system to exist.

Time is based on our use of our planet's rotation and the distances that exist on this planet and outside as our galaxy. But the importance of this fact is that this rotational spin was placed on matter by ***IT*** when ***IT*** threw ***ITS*** weight outwards. This is why we say that it was after the moment of the Big Bang that we could apply time in relation to measuring the distance between the

portions of ***ITS*** fragmented weight as it now exists, and the spin that began existing after the Big Bang. It is important that we remember that before the Big Bang ***ITS*** weight did not have distance because it was all concentrated into a singular point, and we do need distance in order to apply our mechanical time system.

Furthermore, we are programmed to see ***ITS*** weight in our existing moment as something that can be reshaped into something else, in time. However, you will see that ***ITS*** weight that exists as this existing moment will be the same existing moment that ***IT*** has always existed in, that we call the future.

❁~~~~~~❁

*** ***The things we have to do are not consuming our time; they are consuming our existing moment*** ***

⌘~~~~~~~~~~~~~~~~~~~~~~~~~~~~~~~~~~~⌘

Time can get you cross-eyed

Let me explain this, but before I start this piece, I want you, the reader, to remember that this is just a joking way of saying something.

So let us say that you are looking to the past, and to the future, and that as you look to the past you use one eye, and if you use the other eye to look into the future, you could end up cross-eyed!

The best way to exist is to use your line of vision to see things as being in front of you, for the existing moment, since we are made to see best when we are looking straight ahead at what exists.

Let me give you an example: When I visit a museum that has an artifact of the past, I know that what I am seeing as the object is not in the past. The object is also in this same existing moment, for it too as an object cannot exist outside of the existing moment, and the reason why the object is still here in the museum is because the object has resisted reshaping into something else as matter, and if the object is made of something like calcium which can resist the heat from the Sun, it will stay around for more of Earth's rotations, so that when someone that is born thousands of

rotations later, as the same existing moment, will be able to see how ***IT*** existed as a possibility.

We can see the past

The reason why we can see the past is because we can see how ***IT*** existed through ***ITS*** reshaping of ***ITSELF***, for ***IT*** has left us with memories in our minds, and as photos, and as skeletons, just to mention a few. But the reason why we cannot see the future is because ***IT*** has not yet reshaped from this moment into something else, so we do not have anything to confirm what ***IT*** has not yet reshaped into.

Another reason why we cannot see the future as we understand it is because it is ***ITS*** weight that will have to change from how it exists now into what it will change into, as this existing moment that ***IT*** exists as, known as omnipresent.

In addition, let me mention that when we say that things are falling apart in our time system, they are not, for it is just that ***IT*** is reshaping ***ITSELF*** as ***ITS*** weight into something else.

Time and Omnipresent

In order to see and understand ***IT***, or pure energy, better, I suggest that you stop thinking or seeing things as time and just see ***IT*** as omnipresent, as just a moment of ***ITS*** existence in which we have been given the gift of participating in ***ITSELF***.

Once again, let us return to the photo of the young and old lady, and remember that both are there, just as time is here as our convenience, and simultaneously as ***IT*** existing as something we can just call an existing moment (place) where ***IT*** exists.

And remember that as we focus on ***ITS*** duality we will grow mentally, for we will be exercising our minds more, and best of all, because we will see and understand ***IT*** better in ***ITS*** way of existing as a duality. Nevertheless, however much I may describe ***IT*** to you, there are things in our life that we will have to learn on our own, such as finding our own connection with ***IT*** only.

I have found that this invention that we have put together for our convenience called TIME can be a major obstacle in our trying to understand ***IT*** better. This is why monks and priests leave out time when they meditate on ***IT***. Yet I know that there are some minds that would like to tell ***IT*** what time ***IT*** should be there for us.

I am also aware that it was these monks that used our planet's rotation to start their meditation, getting up to pray and meditate just before our Sun started to shed ***ITS*** light, (***ITS*** weight) and I am going to venture offering a reason as to why these monks started their prayers so early. I feel it is the same reason why I also do it, namely, that I prefer to connect to ***IT*** first, because my mind, which was resting from this outside world during sleep, as I return to my outside world, it immediately wants to attach itself to ***IT*** as ***ITS*** weight, that is, to the material objects that exist outside of us.

So I find that if I stay as close to connecting to ***IT*** first in meditation, it is easier, because my mind has just returned from being inside of me during sleep. Let me also add a personal note: I have found that as our minds, we do not DREAM of ***IT*** as ***ITS*** nothingness. I feel the reason for this may be that in dreams what we are seeing are images that the mind put together, since our mind cannot stop functioning, even during sleep! Therefore, it is hard for our minds to dream of ***ITS*** nothingness, for even if we could dream of ***ITS*** nothingness our minds would not even know that ***IT*** is there. The mind cannot even understand ***ITS*** nothingness, even if ***IT*** is there during meditation, and also, because ***IT*** is there as ***ITS*** nothingness in every atom of our very bodies.

Now, let me explain why it is better to start my awakening with connecting with ***IT***: When I wake up, I immediately connect to ***IT***, before I do anything else, because in my personal experience I find that otherwise my mind wants to focus on the situations that are happening outside of me. My attention is diverted and connecting to ***IT*** will be more difficult.

There are many writers out there that prefer to start writing late at night, and go on into the early hours of the morning. But that does not work for me. I have found that after I connect with ***IT***, and I have had a few cups of coffee, I can then start working on the manuscript of this book, (from 4-5 AM). However, if I do get involved with something else, it is hard to come back to writing again because the moment that I do get involved with something else as ***ITS*** weight (I am referring to material things), it does have a pulling effect that makes it harder to tear away from how ***IT*** exists outside of me as ***ITS*** weight.

Time systems

All time systems exist in the same existing moment as omnipresent. Let me explain this with a simple example:

Let us say that there is a person (who to me is a living, existing moment) called Tom who lives in California, it is 11.30 PM, Sunday, and he is about to go to sleep.

And I should also mention that Tom has an exact twin brother that was born at the same moment of life, who was given the name of David, who lives on the opposite side of the continent, in New York. For David in New York, it is 5:00 AM, Monday, and since David is an early riser, he is preparing to go to work. But something happens that requires that Tom, at 11:30 PM Sunday, in California, to call his twin brother David in New York, for whom it is 5:00 AM, on Monday. I am using this example to show you, the reader, why our mechanical time system is not universal. Being twins, they started their human lives at the same time, but now each one has adjusted to live in different time. However, regardless of the hour or the date, when Tom makes his emergency call to David, while they are both are talking on the telephone, they both exist in the same living, existing moment because both time zones exist within the same omnipresence.

Now you can understand why I insist that all mechanical time systems exist in the same existing moment as omnipresent. Even more, let us say that Earth did not exist and both brothers where

somewhere else inside this Universe, regardless of where they might be, they would still be existing in the same living existing moment that they have been permitted to exist within ***ITSELF***, as ***ITS*** existing moment that we call omnipresence.

Why there is no past or future

Here is one more reason why we cannot return to the past, because to do this we would have to UNrotate this planet's spinning, and if we tried to go into the future, we would have to make our planet make the necessary rotations that the planet has not yet made, for our planet, like us has to do things only as one existing moment.

The past and future are omnipresent

Here is something to think about: For a past or a future to exist, this past or future would have to exist within ***ITS*** omnipresent moment of existence.

Time as omnipresent

I will try to begin this subject by referring to the information that is available concerning time.

The evidence presented in this book has led us to the inescapable conclusion that GOD (***IT***) is every thing, and is in all places at the same moment, even our own existence on a planet we call Earth, located in a galaxy that we know as The Milky Way, the size of which alone is estimated to be approximately 100 thousand light years across. I have already mentioned elsewhere that everything that exists in this Universe is really just ***ITS*** weight reshaped into matter. Now, since everything that exists is ***IT,*** as ***ITS*** weight, and ***IT*** is in all places at the same time, if you look closely you will see that ***ITS*** weight is fragmented and distributed throughout this place that we call the Universe, but each fragment of weight is obviously distinct. For instance, the way ***ITS*** weight exists as this planet called Earth is not and cannot be the very same weight of other planets elsewhere, let's say at the other "end" of this Universe. Therefore, it is natural to think of there being distance from one fragment of weight to another within this Universe.

However, even though all matter is just ***ITS*** fragmented weight that exists within ***ITSELF*** as this Universe, we have become attached to thinking of ***ITS*** weight in terms of matter and time. This came about because we applied the concept of time to ***ITS*** weight in order to be able to calculate the distance from one fragment of ***ITS*** weight to another. Nevertheless, this does not affect ***IT. IT*** continues to be in all places at the same moment because ***ITS*** weight is to be found within ***ITS*** cold, clear, invisible "shell body". It is we who have invented and superimposed the concept of time onto ***IT***. However, it is within ***ITS*** nothingness where everything else exists and it is ***ITS*** nothingness that is in all places at the same time, be it as the 95% nothingness that we call this Universe or the nothingness that is found inside every atom; for these two are one and the same.

So time cannot be applied to this nothing that ***IT*** exists as, for how can we apply time to something that exists as a form of nothing?

ITS timeless body

Here is a concept that, having read the previous sections, you will better understand. It has to do with the existing vocabulary that deals with the way ***IT*** exists as omnipresent, and the way our current language is only made to deal with ***ITS*** weight as the things to which we can apply our time system.

I will try to explain this the best I can, but you, the reader, must remember that I too have to use the vocabulary that now exists, which does not have many words that I can use to express ***ITS*** existence as only one existing moment. Do you see? In the previous sentence I had to use the word moment, but I am not using it as a time reference.

For most of humanity's existence we have been thinking in terms of time, yet ***IT*** does not exist as the time system that we have put together for our convenience. The problem is that we have gotten so used to thinking this way that we forget that we have only been referring to ***ITS*** weight as affected by time, such as when we say that the Earth or the stars are millions or billions of years old, even when we know they exist within ***ITS*** timeless body.

If you give this some thought, you will realize that ***IT*** has always existed, but not as time nor affected by what we call time, because when ***IT*** existed before the Big Bang, where ***IT*** had all ***ITS*** weight in one place, our time system did not even exist yet.

So let me discuss now the fact that ***IT*** exists as just existing, and not as even one moment, for I still do not have a word that can be used to describe ***IT*** as just existing, as being timeless, in a "place" where time does not exist. Once again I request that you use your imagination to try to see how ***IT*** exists, where everything that is happening, is happening within ***ITSELF***. This was so, long before ***IT*** reshaped into the matter that now exists, to which we apply our time system. You may have noticed that in the previous sentence I had to use the word "before" and in the same way, I had to use the concept of a "later" moment to describe how ***IT*** reshaped into matter! All this difficulty is caused by our now existing vocabulary that is based mostly on things that have been (the past) or will be (the future).

Continuing to try to describe the way ***IT*** exists as just existing, not needing to use time to reshape, for want of a better word, I will use the word "place" in order to avoid using words that have to do with our time system. However, you, the reader, must be clear that I am not referring to a physical place inside which ***IT*** exists, for ***IT*** cannot exist somewhere other than within ***ITSELF***. When I say a "place", I am really speaking of the place that is ***ITS*** body as the nothingness that now exists as this empty Universe, inside which ***ITS*** weight exists. Here we can apply time to the things that exist inside of ***ITS*** nothingness, as ***ITS*** weight, which ***IT*** also exists as, inside of ***ITS*** transparent body, which is made from a form of a nothingness. Of course, we cannot explain this way that ***IT*** exists because there is nothing there for us to describe as existing. For this reason we cannot even use our time system to describe the way ***IT*** exists as this clear, transparent way that ***ITS*** timeless body exists. And it is within this transparent nothingness that ***ITS*** fragmented weight is moving about. But here we cannot apply our time system to this transparent nothingness nor to the way ***ITS*** weight exists inside of ***ITSELF*** as this timeless clear nothingness.

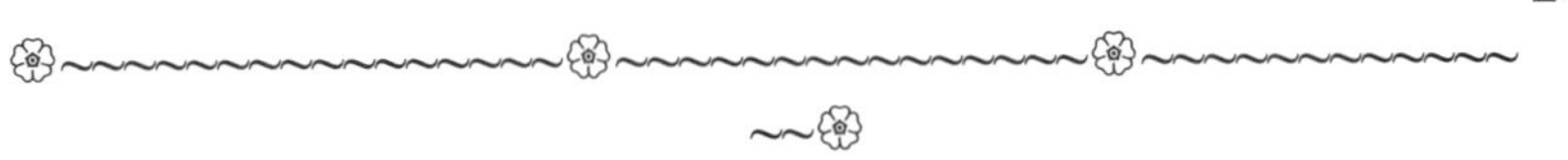

******Our vocabulary is the first thing that makes us feel old. We are introduced to the concept of age or "getting old" as soon as someone tells us that we are one year old.******

So, for lack of a better way to say it, it was "after" ***IT*** fragmented its weight, that ***IT*** allowed us to exist as ***IT***, within ***ITS*** clear, transparent body; a lengthy process that "started" with what we call the moment of the Big Bang, that led up to the existence of our planet and the emergence of humanity upon it.

Once again, when we speak of the Big Bang, we have to make reference to our time system, so that we can better understand, that before the Big Bang ***IT*** existed as one singular weight, and then this weight became fragmented into all the contents of what we know as this Universe, that are moving about within ***ITS*** clear timeless body.

This is the only way we have, as far as words go, to understand something about the way ***IT*** exists. We have described the way ***IT*** existed before the Big Bang (the past), and we can talk about the way ***IT*** now exists, as the way ***IT*** reshaped ***ITS*** weight (present), or the way ***IT*** will reshape ***ITS*** heated weight into what we call the future, for we have no words that describe ***IT*** as just one continuous way of existing, where ***IT*** does not have a past or future, where ***IT*** only exists as this omnipresent. What word can I use to say "moment" without using the word moment as time??

******IT will not punch a time clock to satisfy us. ******

Now since we are made in ***ITS*** own image, as the way ***IT*** exists, let me now try to describe the way we exist, in the same way that ***IT*** exists, but without reference to our time system.

Now that we have talked about the ways that ***ITS*** weight has been changing, which is where we began to apply time, let us stop to just look at ***IT*** as the way ***IT*** exists, where ***IT*** is reshaping ***ITS***

heated weight within ***ITS*** cold, clear, transparent body. This is where ***ITS*** weight exists, just moving about within ***ITS*** cold, clear, transparent way of existing, which could be called ***ITS*** body. And it is here where ***IT*** is continuously reshaping into something else as ***ITS*** fragmented weight. In one of these possibilities into which ***IT*** reshaped ***ITS*** weight, we came into existence, as ***ITS*** weight, as ***ITS*** life, and as ***ITS*** conscious divineness, for everything we exist as is because ***IT*** exists that way as ***ITSELF,*** as just one, that we call GOD, or as the pure energy that exists as this cold, clear Universe, where we know that ***ITS*** heated weight exists, where we have confirmed that **IT** does exist , and we have confirmed that we do not know what ***IT*** is, but we can confirm the way ***ITS*** weight has behaved which we refer to as transmutation. This is the way ***ITS*** heated weight behaves within ***ITS*** cold, transparent, timeless nothingness.

Now if we forget about using our time system and we look at everything that is happening as happening only to ***ITSELF*** as ***ITS*** now fragmented weight, all of which is happening within ***ITS*** cold, clear nothingness, to which we cannot even apply time we might begin to understand the concept of omnipresence, which is the fact that ***IT*** is in all places as the same moment, and if we remove the word moment, we can restate the concept as follows: ***IT*** is in all places as ***ITS*** nothingness, where not even our mechanical time system can be applied, for we will never be able to apply time to how ***IT*** exists as ***ITS*** nothingness, and this is also why we cannot apply time to the way ***IT*** exists in what we call the activity that is taking place inside the atom, known as subatomic activity, where ***ITS*** fragmented weight also exists.

Another way to understand this which has to do with time is by visualizing an atom. The fragmented weight that makes up the inside of an atom is part of ***ITS*** total weight and this fragmented weight that exists inside of an atom can only exist within the empty nothingness that ***IT*** exists as and the fragmented weight that exists inside every atom came from when ***IT*** had all ***ITS*** weight in one particular place inside ***ITS*** nothingness “before” the Big Bang. “After” the Big Bang ***IT*** threw ***ITS*** heated weight outward within

ITSELF as ***ITS*** clear nothingness, which can be regarded as ***ITS*** "body". It is important to remember that ***ITS*** weight exists within ***ITSELF***, not outside of ***ITSELF***, because this would mean that wherever or whatever this "outside" place is would be independent from ***IT***, as the pure energy that now exists inside this Universe as ***ITS*** nothingness.

Another thought, related to the one above, has to do with why we cannot get a mechanical clock inside an atom. One reason for this is that we use the rotation of our planet as the basis of our time system and this does not exist inside an atom. The other reason is that we use our time system to measure what ***IT*** does with ***ITS*** weight out here where we can exist.

For us it is easier to understand what is happening out here where we are, as ***ITS*** fragmented weight, compared to trying to understand ***ITS*** also fragmented weight that exists inside atoms, as smaller portions of ***ITS*** fragmented weight. However, we must not forget that both ***ITS*** outside fragmented weight and ***ITS*** fragmented weight inside the atom are existing or happening in what we call omnipresent, which is how everything that exists, exists as ***IT***, as just being ***ITSELF,*** as one.

Please keep in mind at all times during this discussion that we are really just referring to ***ITS*** weight that exists both outside of us, as well as inside the atom. Both exist within ***IT***, as ***ITSELF***.

The time system that we put together for our convenience is great when we have to be in at a particular place on this planet at what we refer to as "a certain time". Nevertheless, whatever we choose to call it, whatever time bound words we use, we are still speaking about the same particular moment that we exist as, which is also going to be the same particular moment that exists wherever we may go inside this Universe, as the way ***IT*** exists.

I would also like to clarify that when I speak of ***ITS*** nothingness, it is not because there is nothing in this nothingness, for there is!

This is where ***ITS*** high speed (***MAXX-SPEED***) exists. The problem is that this speed is faster than the speed of light, because as light, ***IT*** places a minute amount of ***ITS*** weight as the heat that light carries, but as ***MAXX-SPEED,*** there is nothing there as ***ITS*** weight that our sense could see or detect. The other reason why this area is timeless is because this area is faster than the speed of light, which is the speed at which time stops naturally. In any case our time system can never be applied to ***ITS*** high speeded nothingness as the way that ***ITS*** clear transparent body exists.

Knowing this, you can look at this Universe as a timeless place, which is really ***ITS*** body, where ***ITS*** weight is moving about, as this omnipresent place that exists as ***ITS*** clear, transparent form of being. It is evident that everything is happening to ***ITSELF*** as ***ITS*** weight that is moving about, as this weight transforms (reshapes) within an area where our mechanical time cannot exist, which is something we can deduce from the fact that ***ITS*** weight was all in just one place before the Big Bang. During this phase our mechanical time system could not be applied to ***ITS*** weight, firstly, because we did not exist, and secondly, because we need ***ITS*** weight as matter in order to apply our mechanical time system to this weight as it moves about.

It should be clear to you, the reader, by now that it is ***ITS*** weight to which we apply our mechanical time system, but ***ITS*** weight exists within ***ITS*** timeless way of existing. Everything that is moving is doing so within ***ITS*** timeless way of existing, as just a place where ***ITS*** weight can reshape. This timeless ***SELF*** which consists of ***ITS*** fragmented weight moving about inside ***ITS*** nothingness is what we call the Universe. And we cannot exist outside of this timeless nothingness, because we also exist as ***ITS*** weight, which can only exist within ***ITS*** timeless body.

Now you will realize that even if we tried to apply our mechanical time system to ***ITS*** weight as a whole we would not be able to, because we and everything else are really just fragments of ***ITS*** weight that are moving about within ***ITS*** timeless body.

Everything that exists is part and parcel, and exists within this omnipresent body that is ***IT***. ***IT*** is timeless, ***IT*** is omnipresent.

Knowing this you will now see why we can only say that we exist as.... Here again I cannot find a word to describe existing as just existing, without any reference to our mechanical time system, so I hope that there is a reader out there who will find words that we, as humans, can use to describe the way we exist within ***ITSELF*** as a timeless place (body). Here again I am caught having to apply the idea of a future time, when we would have as this same existing moment found ways and words to talk about how we as ***IT*** exists in this timeless nothingness, within which we exist with ***ITSELF***. So just keep in mind that even though I have to use the language of time, I am not being inconsistent. When I use the word later, referring to "the future", this means "with the passing of many Earth rotations".

As much as I have tried, I feel I have not done well, for this way of presenting things in words is the best that I could do under my circumstances or situation, because these are the only words I know.

Nevertheless, my aim is to tell you, the reader, how you can become aware that you are living in the only existing moment, where there is no past or future, for we do not know how ***IT*** will reshape ***ITS*** weight within ***ITS*** nothingness. To accomplish this, you must keep in mind that the rotation of our planet, (which is a portion of ***ITS*** weight existing within ***ITS*** nothingness) does not help us to see that during all of Earth's rotations, we have always been within one same, living, existing moment that ***IT*** permitted us to exist in, as **ITS** weight, within ***ITS*** timeless nothingness, which is ***ITS*** timeless way of existing, as the way ***IT*** exists.

Let's try another thought experiment: Think about anything that has happened to you (in "the past") and you will notice that everything that has happened to you during your life as happened as the same living existing moment that you now exist in, and

knowing this, you will see that anything that will happen to you during your existence will also have to happen as this same living existing moment in which you are reading this page. If you hold on to this you will be closer to *knowing*. I say knowing, because you have always existed as ***ITS*** TIMELESS existence as ***ITS*** weight as matter which can only exist within ***ITS*** nothingness, as the nothingness that you exist as, your body being made up of atoms, and the way these atoms also have to exist within ***ITS*** timeless, clear, transparent body, that we refer to as omnipresence.

For we have always existed within ***ITSELF*** as ***ITS*** weight within ***ITS*** timeless body, as just a way of ***ITS*** existence, not governed by our time system. And the same way ***IT*** has always been reshaping ***ITS*** weight within ***ITS*** timeless body, is the same way we exist, in ***ITS*** own image, in the way ***IT*** exists as just being timeless, for ***IT*** never had a beginning and ***IT*** will never have an end. I am very grateful that ***IT*** exists this way, for this is what will permit anything that will happen to happen as the way ***IT*** will hopefully continue to reshape ***ITS*** heated weight, within ***ITS*** cold, clear, transparent body, into infinite possibilities.

Perhaps we can wait until someone comes along and writes a novel where everything that is taking place in the book is happening as just one moment (omnipresent). To do this, the writer would have to think of all events as just happening to ***ITS*** weight, as just moving about within ***ITS*** TIMELESS body, as events that are happening in this place called omnipresent, but without reference to our human mechanical time system.

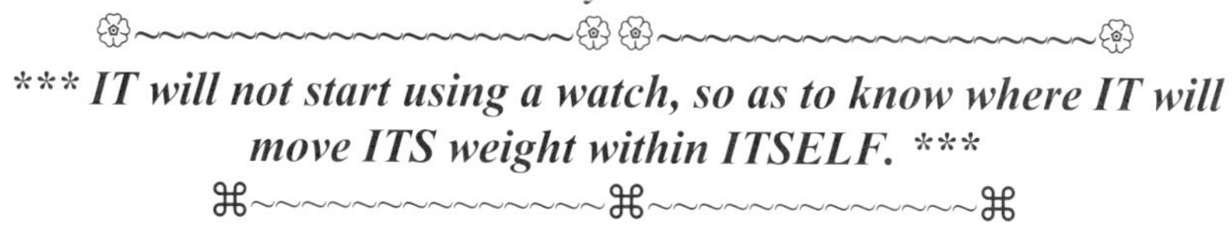

*** ***IT will not start using a watch, so as to know where IT will move ITS weight within ITSELF.*** ***

IT is more

When I have tried to apply time to ***ITS*** existence, wondering when ***IT*** came into existence, I had a problem finding matter to which I could attach time as a starting point in ***ITS*** existence. For example, we refer to this Universe as having been existing so many billions of years, but this means that, as time, this Universe started to exist

many billions of years ago. When we say this we are forgetting that the Universe exists as pure energy that has no beginning or end, and that pure energy is also God.

If I try to see ***IT*** as having a beginning, it does not work because for us to attach time to this Universe is the same as trying to attach time to ***IT***. So, I stopped thinking as if ***IT*** had a past and future, and stayed with ***IT*** as only an existing moment. I tried looking at the information that has been orally transmitted for generations relating to ***IT*** as God and found that we have many interpretations of God that include many reasons concerning why a certain road is the proper road. There are also many names for how God exists, even some related to ***IT*** being a God that only exists for us here on this planet. Some interpretations refer to God's positive and negative ways of doing things, but most neglect to refer to God as a singular and omnipresent God that exists as each single atom or that exists as the other side of this Universe as well. Few mention ***IT*** as a duality.

After reviewing the religious information that is available concerning ***IT***, I started to connect the scientific information that was available showing that ***IT*** is pure energy. This information indicates that ***IT***, as pure energy, has no beginning or end, for ***IT***, as pure energy, just exists. ***IT*** exists as a positive and negative force. ***ITS*** positive is ***ITS*** mass, or weight. This has been tracked down to when ***ITS*** weight existed as that very dense matter just before the Big Bang. ***ITS*** positive part is still here, as this moment, as the mass in every atom.
All of this mass, which was once considered to be 10% of the Universe, but we now know, with more accurate measurements, to be closer to 4.6% of the Universe, exists as a positive. I would like to add that when we refer to ***IT***, as pure energy, being a positive and a negative, we should remember that these are just words that the human mind has invented to help us understand our existence. We are the ones that state that when something is positive it is good, and that something negative is bad. We should always remember that the words we use are there to help us understand our existence.

I have said all of the above because if I use these words that we are accustomed to, as to something being positive or negative, it will mean to our understanding, that ***IT***, as God, as pure energy, exists as being from 4.6 to 10% positive. This is all the positive energy that exists in every atom within this stage of the Universe that we are now in. The opposite is the other force of pure energy- the 90 to 95.4 % nothingness that is the negative force of the Universe.

❁~~~~~~~~~~~~❁~~~~~~~~~~~~~❁

Memories are what IT uses to remember what IT reshaped from, or how ITS weight used to be or how ITS weight once existed.

⌘~~~~~~~~~~~~~~⌘~~~~~~~~~~~~~~⌘~~~~~~~~~~~~~~⌘

If the 4.6-10% is composed of what we call a positive energy that exists as the heated weight that is matter, then we must take the opposite of the above, which would mean that the Universe is 90-95.4% negative. So, you will understand why ***IT***, as God, as pure energy, is more negative than positive, at least to our human understanding of things, to which there is no problem, for ***IT*** will continue to exist the way it does, not the way we think it should.

IT will continue to exist as one, as what we can call a *moment*, which is also just a word with which to say that something ***is***. Since we have built up so much information on ***IT***, as this pure energy's positive and negative, we should always remember that what we are calling pure energy, as what is positive, will always be ***ITS*** heated weight. Also, what we have been calling this pure energy's negative, ***ITS*** cold, clear nothingness that exists as ***ITS*** invisible shell is what we are now calling *this Universe*, the pure energy that has always existed as ***ITS*** duality.
And now that we are more educated, we will be able to adjust to seeing and understanding this negative that exists as God, ***IT***, as a cold, clear nothingness that does exist. Imagine if we tried to wipe out this clear, cold nothingness that is the negative force of the Universe; we would not have a place for the positive part of ***ITSELF***, as pure energy, to exist.

❁~~~~~~~~~~❁~~~~~~~~~~❁~~~~~~~~~~❁~~~~~~~~~~~❁

*** ***We exist as an illusion, because we are not we, we are IT in ITS totality.*** ***

The longest ending

I would like to end this book by asking everyone's pardon for the way I put it together. This is my first attempt at writing. I never would have considered publishing a book before, for various reasons that I would like to share with you.

I grew up in the major ghettoes or what are known as hot spots in Brooklyn and Manhattan, New York City, and in Puerto Rico. I dropped out of school at the age of sixteen, which was the legal age at that time. As a result I never was able to get a grip on spelling, grammar, or any other aspects of writing. This is why there are no hand written love letters from me to any of my old girlfriends; no one would have understood what I would have written.

And for those of you that live in New York City, here are some of the places where I have lived in Brooklyn: 25 Siegel Street, 29 Moore Street, 50 Howard Ave., 121 Amboy St, 169 Stone Ave. I am sure that what exists at these addresses today are housing projects, for this has been the trend for New York, since the city's population is increasing. Building upwards gives people more room in which to live and is cheaper, since real estate is becoming very expensive while "air space", so to speak, is free.

The first work that I performed for money was shining shoes. Then I started working in factories and worked my way up to selling in stores.

I have to be thankful to certain people when it comes to my education. It was Bob Henderson who introduced me to the wonderful world of science fiction books. This started my love affair with reading, which changed my life by opening the Universe to me. I am still inspired by what Bob said to me once; he said that I had an over-rational mind which sponged-up everything I was exposed to. Before this, all I ever read were comic books and Mad Magazine, which had plenty of photos and very little writing. My favorite section was Spy vs. Spy.

I finally left off doing indoor work, when by luck, or rather, when ***IT*** sent George Mercado, of Pavarini Construction; who, seeing that I could be of use to him (as cheap labor), took me under his wing in the engineering department. There I learned about engineering and construction, which gave me insight about how to put things together, which has been an important occupation for me throughout my life.

But getting back to my not being able to write and how this book got done, even now I am not clear on the definitions of vowels, nouns, or even where the comma should be put; not to mention all of the other rules. Then a friend and computer expert, José Castro, built me a computer from old parts and showed me how to use word processing softwarc; especially features like spell-check. I also started using Dragon Naturally Speaking, which made it possible to play back (text to speech) so that I could hear what I was writing. I also learned the importance of the save button and how to be very careful with that disastrous delete button.

As for computers, I believe that somebody could become richer than Bill Gates, if they could develop a hammer or somcthing that a user could show to the computer to scare it. Something so that when it gives me a hard time I can show the computer and say "You either shape up or I will tune you up with this!"

Let me also say that I never had any intention of writing a book. The idea came from a friend. I had shared a view on something with him and he turned to me and said that I see and say things in a strange way. He said I should try putting the things I was thinking about on paper so that he could read them later. But many of my friends did not know how little formal education I had. My degree is from the streets and hard work. So I tried finding someone that could transcribe my views using a pre-recorded audio cassette tape of what I was trying to say. This did not work because, not being properly educated, I was not talking properly nor clearly enough for someone to listen and interpret. After a few attempts I saw that this was not the way.

Since I was not doing this out of need, or because I was looking for financial gain or personal recognition, it was a slow start. If it were not for computers becoming cheaper, this manuscript would have never gotten started or finished. What began as notes on different subjects grew to a point where a friend asked if I had considered putting them together in a book. That's when I started giving the idea serious consideration. I told my friend that I did not think this could be a book because I did not have many pages of material. I was told it is not the number of pages that mattered but the content. Books have been written with 500 pages when 100 would have been enough.

I was told by my friends that they would buy the book if I published it; not for the size but for the content. I was also reminded that there was a lot of good information that should be shared, and that I should leave the door open for communication with others that would like to add to the dialogue.

Let me explain my experience with the computer. After my friend José gave me a few classes on how to use the keyboard, spell check, and how to look up different words, I started to type. I have included Samples 1-5 below so that you can see my progress in bettering my composition. Of course, I am no typist; I just pick and peck my way around with two fingers. Anyway, I found it faster to use the keyboard and correct my misspelling with Windows 98 spell check and grammar check. Then I would listen to what I put together using the Dragon Naturally Speaking text to speech program, and redo what needed corrections until I got it close enough to where one of my ghost writers could take over, which is what you are now reading.

After many of earth's rotations I got a little better at picking my words. If I was not sure of a word I would look it up in the dictionary. I got many words from spell check. I would work to get it close enough for the ghost writer to redo it so that the text would say what I intended, without them ever knowing the truth about how bad I was in my literacy. That is what happened with

Diana Villafañe, who was one of the last ghost writers to work with me,. She never knew the truth about my literacy until I gave her this section at the very end, as the ending to this book. And as soon as I saw that employing a ghost writer would work, I knew that I had found a way to having my thoughts put on paper for others to read.

However, this meant I would be the one to sit down and have to put my thought into words, but since my beginning of anything that I write about looks more like hieroglyphics, I knew that I was going to have to work, in changing my first draft of hieroglyphic writing into something a ghost writer could find clear enough to work with, and end up with what you are now reading. Please believe me when I tell you that I sometimes rewrote sentences or paragraphs more than 5 times. But this was my new job, and ***IT*** was not rushing me. Actually the longer it took me to write something the more I kept learning about the way ***IT*** exists.

By the way, the hardest part of this project has been pecking away at the keyboard and trying to find the right words. I knew it would be difficult as it was something I had never done before. But I did my best. I just have to ask ***IT*** to forgive me for not portraying ***IT*** properly in my plain, uneducated, blue collar, ghetto style.

As we kept working together I saw that the ghost writer understood what I was trying to write. A lot of the subjects are a bit off the wall or somewhat crazy. I had to find every available way to make my thoughts as clear as possible so that the ghost writer could get my point and express it better. Many times I just kept saying the same thing in different ways to the point of redundancy. All of this has shown me that I do better at expressing my ideas by not talking so much and just writing them.

Throughout the process the ghost writer has had a lot of patience with my original manuscripts. They have known how to get into my head through what they read from me and they have been able to translate my thoughts into readable form; certainly much better than I could have. The price I have had to pay is having to accept

how ghost writers reduce a lot of my writing into just a couple of sentences.
Let me share with you something that I learned about writing. You will understand better if I begin by saying that when you start writing a letter you already knew more or less what you are going to write about. So when you start writing it is easy just to present a subject on about which you already have information. But this is not what happened when I started writing. All I had was just enough information on just one particular topic, for instance, contamination, so I would have to search for words that the ghost writer could use, so that my text would become what you have read in the piece called Contamination. This is why this book started out as a collection of Trivia, which were just small sections on different topics, so that by the time I was mastering writing with these small sections, ***IT*** prepared me, for what came next, starting with part #2 , which I never knew existed in me when I began. I had no idea this was ever going to go beyond section #2 and into something that I never knew anything about, as in section #3 which is on ***ITS*** cold, clear, transparent nothingness, or that in writing this manuscript I would have to change the way I understood time! But before I knew it, as I continued writing I wound up writing about things that I had not started out to write about.

I find this interesting enough to pass on to those that are interested in writing, but feel that they don't have much to write about. I more or less felt that way when I began, but I now find myself continuously writing about things as they keep coming into my head. Let me mention that now, when I get a thought in my head, which at times comes in the middle of the night, I first use a cassette tape to record it, so that by re-hearing what I recorded I can begin to re-construct the subject or topic.

I have had to stop adding more to this manuscript so that at least you could read what I have found until now, about ***IT*** and its many ways of existing. So, my advice to those of you who want to write is: Just start!

If I can do it you can too

Another thing I like about working with a computer is that I can see when I have a misspelled word, because the program underlines it in color. This also helps me know what portions I have not yet completed. After I correct the spelling and the colored underlining disappears, I can begin to listen to what I was trying to write about. Another thing that has helped me is putting together a list of spelling words, that contains those words the spelling of which I have the hardest time remembering. Now, instead of using the spell check program on the computer to offer other words, I just look at the printed copy of my most problematical words which I keep under the first page of my dictionary. I must also say that it does help to use a dictionary, for after the computer gave me the right way to spell a word, I would then look in my dictionary to check out to make sure that the definition was what I was looking for, and I have used my dictionary so much that it is now falling apart!

Get ready now, dear reader, to witness for yourself what my hieroglyphic writing looks like before I send it to my ghost writer. Below I have included 5 samples that will progress from my first words to what the text looks like before I send it off to the ghost writer to be cleaned up and, as I say, “repaired”.

Here is how time and memory got started, in 5 samples, and a portion of how it ended. I am including my way of writing so that those of you who think that you will never be able to write will just remember that I doubt if you write worse than I do!

Now look at the fallowing pages as my best writing.

SAMPLE #1 as the beginning of Time and Memory

We have experieniced what we call an event from the past, existing

Here is an other way of understaniding what we call the past, that exist as this existing moment

Maybe this happened to you.

So that you can see , that what happined as an event , did really happen as this same moment that you are now livind in , and this is why you feel , that you can rememimber this event , as if it hapapen right now , as the moment you are now in, because in reality, it did happen , as this same moment , but many rotatons ago , because when it happaed then ,(the event) it happenned in this same living exising moment , but since what ever the event was , you will not be able to see it , the way it happened , becasuse that event, has alredy changed . and you can feel , and the reason why you can remember that event that took place many rotations ago , as is it happened just now, is because that moment, and this moment are the same moment , but since that event changed , all you have now is the memoreie

SAMPLE # 2 time and memory this is after my listening to it on dangon

We have experienced what we call an event from the past, existing

Here is an other way of understanding what we call the past, that exist as this existing moment

Maybe this happened to you.

So that you can see , that what happened as an event , did really happen as this same moment that you are now living in , and this is why you feel , that you can remember this event , as if this is hapapening right now , as the moment you are now in, because in reality, it did happen , as this same moment , but many rotations ago , because when it happened then ,(the event) it happened in this same living existing moment , but since what ever the event was , you will not be able to see it , the way it happened , because that event, has already changed . And you can feel, and the reson why you can remember that event that took place many rotations ago, as is it happened just now, is because that moment, and this moment are the same moment, but since that event changed, all

you have now is the memories

SAMPLE # 3 time and memory

this is after my listening to it on dangon the thried time

We have experienced what we call an event from the past, existing

Here is an other way of understanding what we call the past, that exist as this existing moment

Maybe this happened to you.

So that you can see , that what happened as an event , did really happen as this same moment that you are now living in , and this is why you feel , that you can remember this event , as if this is hapapening right now , as the moment you are now in, because in reality, it did happen , as this same moment , but many rotations ago , because when it happened then ,(the event) it happened in this same living existing moment , but since what ever the event was , you will not be able to see it , the way it happened , because that event, has already changed . And you can feel, and the reson why you can remember that event that took place many rotations ago, as is it happened just now, is because that moment, and this moment are the same moment, but since that event changed, all you have now is the memories

SAMPLE # 4 time and memory

after drogon , but since dragon does not have a good spiel check , I have to do it one more time , so that I can clean it up a little more , before I send it to Ghost writer , so here goes

we have experienced what we call an event that happened in the past, and we say that we can remmber that event as if it just happened is because of this

Here is an other way of understanding what we call the past, that exist as this existing moment

Maybe this happened to you.

So that you can see , that what happened as an event , did really happen as this same moment that you are now living in , and this is why you feel , that you can remember this event , as if this event happened just now , as the moment you are now in, because in reality, it did happen , as this same moment , but many of earths

rotations ago , because when it happened then ,(the event) it happened in this same living existing moment , but since what ever the event was , you will not be able to see it phiscally , the way it happened , because that event, has already changed . And you can feel, and the reson why you can remember that event that took place many rotations ago, as if it happened just now, is because that moment, and this moment are the same moment, but since that event changed, all you have now is the memories of that event

SAMPLE # 5 time and memory

And in 5 is after I return to word poerfevct , and do the final correctiosn , as my best before sending it out to Ghost writer .

When we have experienced what we call an event that happened in the past, and we say that we can remember that event as if it just happened is because of this, Or

Here is an other way of understanding what we call the past, that exist as this existing moment

Maybe this happened to you.

Below is a portion of time and memory as finished.

When we have experienced what we call an event that happened in the past, and we say that we can remember that event as if it just happened, it is because of this: What happened as a past event actually happened as this same moment that you are now living in. This is why you feel that you can remember the event as if it just happened. In reality, it happened as this same moment, but many Earth's rotations ago. It happened in this same present moment but you cannot see it physically the way it happened then, because the event has changed as ***ITS*** weight.

In other words, the reason you cannot truly remember the event that took place many rotations ago as if it happened just now is because that moment and this moment are the same moment, but since that event changed, all you have now is the memory of it.

I take the blame

Let me also say that I take the blame for the way this book turned out. It is the best I could do at this point. I have had to be my own editor, publisher and distributor. And if there is someone out there that can help me re-edit this manuscript, please e-mail me.

For this reason, I ask you, the reader, to overlook my mistakes and to forgive me for the way I have had to arrange this entire book, for this is the first time I have ever done this kind of work involving the arranging of information. In addition, an interesting thing I learned while working on this book project is that before I just used to read a book, but now I truly pay attention to the individual words that the writer has used to project his views concerning the subject at hand. Also, now I know that talking about something and writing about it are not the same. For all of this, I have to say thank you to ***IT*** for sending me people like Jane Mayo, Joe González, Chrissine Ríos, and most of all to Diana Villafañe who did most of the book, and also translated it into Spanish, but more important, for her extensive knowledge of the subjects covered in this book, and her ability to confront what I presented as my information.

In this process I also had to learn what it is to start writing words with meaning, and all about editing, printing, selling, promoting, and shipping. Luckily, ***IT*** only sent me just one book with which to work. I also have to thank ***IT*** for sending me Diana as one of my ghost writers, who knows more than I about physics, who confronting some of my topics, made me aware that I needed to make certain statements clearer. Let it be known to all that I will always welcome anyone who can make suggestions on how to edit this book so that it is better. Although, I do know that as ***IT*** gave me the information that is in this book, ***IT*** did not give it to me in an orderly way, nor did ***IT*** give me all the information related to a single topic all at once. So I was aware that if were to and take, for instance, all the information concerning ***ITS*** nothingness and re-arrange it to eliminate the repetitiousness it would take me many more Earth rotations, which would mean that this book would take even longer to reach you, the reader.

It's funny to think that I have worked with ghost writers for almost a couple of years now and we have never met in person. All communications have been through e-mails and a few telephone calls. Furthermore, I never told the ghost writers about this ending because as soon as I saw that they knew what they ware doing, I had no need to bring up the subject of my education. It was uncalled for because after they returned what I sent them, I would then listen to it through Dragon's text to speech program and it sounded very good. Then I gave it to friends to read and their comments were that I wrote very well. Of course, they did not know that I could never write that way. And yes, I was worried as I worked on this project, for I knew that if this pure energy did not want this information to get out, ***IT*** would wipe me out of existence before I could finish it. I asked myself why I was permitted to see some of the things I have seen. I have small eyes but I can see through things. I am good at taking things apart and putting them back together. I believe the answer is this: **One does not need a great formal education to understand this God known as pure energy.**

Putting this book together was a great learning experience. I now know what a writer has to go through. I have also learned a lot about organizing and expressing my thoughts and ideas, maybe not in the classical way, but in the way I could with the help I received. It was not until I accomplished my first research project, (which, by the way, was about water), that I came in contact with atoms. It was in this way that I found a lot of the information I have shared with you in this book. I received most of my educational information from reading old, used university textbooks and from seeing programs like Discovery and Nova. I also read a many journals and magazines such as Scientific American and Science News and many other books. We should also never forget that there is a lot to learn from just observing ***ITS*** constant reshaping. There is no book or any other source of information that can bring you in contact with ***IT*** other than observing ***IT*** as this existing moment. By the time you read this, ***IT*** has changed. What is important is to feel ***IT*** inside of you, and to observe ***IT*** in ***ITS***

changing totality; from this existing moment as you, to the furthest reaches of the Universe. So, if it were not for all of the things mentioned above, this book would have never been written.

And let me also mention to the reader that when someone tells me that my thoughts are deep or very serious, or that my thoughts are profound, I tell them that thinking that way, with all those descriptions, is just an excuse to not observe for themselves; for I have not yet found a deep or profound thought nor have I ever been spaced out with one. Truthfully, I have never even gotten a headache while thinking of anything. Thinking is stopping to mentally view something inside that you cannot put on paper; and it is not deep, because if it were, I most likely would have gotten lost somewhere in my mind. There is nothing profound; I still think at the same simple pace I always have. One more thing that I have found in me, that has helped me, is that I have never had any demons in my head to occupy me. The only common factor that I too have, is the need to survive and therefore, having to perform gainful work, and now that I can look back, I can see that this also acted in my own favor, for I still do my best, and I still give all I can.

I'd also like everyone to know that I have not had any serious injuries to my head, other than the near miss that came from my mother's flying frying pan when I did something wrong. I also have to be grateful to my parents for being divorced, because I, and my brother, having to live with my mother, and my mother having to find work in order to support us, meant she left us alone to fend for ourselves. We even had to cook our own meals. Furthermore, since my brother was 5 years older than I, he left me to myself, for he wanted to be with older friends. Looking back, I can see that my independence caused me to have to learn to take care of myself and think for myself. This started very early due to not having someone, such as a father or mother at home most of the time to care for me. For example, if I cut my finger, I took myself to a hospital for stitches. If I got a toothache, I took myself to a dentist. I seemed to have learned early that I could do things for myself, for I knew that my mother did not know how to read or write, and she

did not know how to deal with the institutional system. However, the system did respond to my various calls for there help, since they saw that I too was not very well educated.

I have also learned that in talking about something I am just expressing my thoughts and it is harder to organize these thoughts to write them down for they travel at a very high speed, since their origin is energized by something that is traveling near the speed of light. Nevertheless, when I sit down to write a thought I have to take more time to organize it. I truly have to look closely at what I am trying to say so that the words I select are as clear and close as possible to the meaning I originally intended. Another important thing is that I started learning from my thoughts. The process I have described has helped me find additional related information that I already had inside of me or available from outside sources. For example, I had never really had a profound interest in physics or astronomy, so I was never interested in subjects such as black holes. But when I started to observe what ***IT*** did and was doing with ***ITS*** weight in terms of black holes, I came to the conclusion that what is really happening when a black hole sucks in the surrounding matter is that ***IT*** is using this process to return to where ***IT*** began just before the moment of the Big Bang. So I've come to the conclusion that there are definite advantages to writing things down in words because this opens up insights to ***IT***, and in addition, when something is written down one can go back to it for reference or re-read it if it was not at first clear.

Another reason why this book ended up the way it did is because when friends found out that this book was under way, they wanted to photocopy the existing information so that they could at least begin to read what I had written just as interesting information on ***IT***. I told them that if they just held off a little longer they could read it as I wrote it without editing, for I did give this manuscript to five people that were going to help with their professional skills. Since they did not come through, I decided to print it as you are now reading it. I feel that you will enjoy it the way it is now. Also, it was a good thing I did not have to print copies for my friends, which would have turned out to be expensive and bulky. Anyway,

I did try to find a company to professionally edit this manuscript, but I had no luck, so I went ahead and printed it myself, as an independent book publishing that I called “R&R publisher”.

I would also like to clarify that I have not written this book because I want to act as a teacher. I am not here to tell anyone what to do or teach anyone what is right or wrong, but rather just to share with you some things that I have found. I apply these things to myself. When it comes to learning more, I order myself to do so continuously, to add information to what I already have, seeking more possibilities in my ever-changing existence.

My personal experience has taught me that not being able to write doesn’t set any limit on how much one can learn. If you really want to learn you can continue learning until your very last moment of existence.

The new educational system

Talking about learning has led me to reflect on our present educational system. In connection with this we should always remember: ***ITS*** law is change. All things, all creatures, all systems are subject to change, as ***IT*** reshapes into new possibilities. I don’t know what ***IT*** may have in store for the future but a look at current trends may give us some hints.

For instance, we are closer to having real-time virtual classrooms through the use of television and computer monitors with instant communication between the students and the professor no matter where each one of them may be on this planet. Computer technology has reduced the cost of having to purchase books and lectures can be recorded onto CDs for future reference.

I think a system should be found whereby one can find what one is interested in, such as the internet, because when one is interested in something one absorbs more, learns more rapidly. For this reason I feel a free data base on educational information would be of importance.

As for the educational system we now have, we should seek out what does not work, such as having to physically go to a school building or having to attend with other students that distract one's attention from learning what is being presented. It is certainly more economical to stay home and learn instead of paying for transportation and for the maintenance of physical buildings and their infrastructure. Home schooling does work for some, and when one's physical presence is required, such as for a chemistry class or when a test must be taken, places can be set up without too much ado.

Another learning improvement would be something like an index with a list of subjects where one can see what is available. When one finds something of interest one could then go on to video or audio, because some people learn better through seeing, some through hearing, and others benefit from both. Sometimes one can even learn by touch!

Should there be anyone among my readers who would like to help putting together an educational data bank which could be accessed through the Internet, please let me know.

My Feeling

Now you, the reader, can see why I left this section of the book for the end, so that you would not judge me by the educational diplomas that I do not have and I feel will never need because ***IT*** does not give out paper diplomas. What ***IT*** does give is a personal, inner, peaceful feeling.

This section is about my personal experience. I am not qualified to speak or opine about other people's experience, but I do wish to offer my own because I believe there may be some readers who have had similar experiences. There are many types of feelings we have experienced from childhood through maturity; feelings such as anger, hate, desire for having things, a doll, a bicycle, a house, a car… Sometimes it is not easy to put feelings into words that will give you that feeling of the feeling itself.

To begin speaking of the feeling of loss, I would like you to think back to when you have lost something, and we know that whatever this that we have lost may be, it does make a difference, for when we find it, or it is returned to us, we do feel better, more fulfilled, more at peace. However, there are losses that cannot be assuaged because there is no going back, such as the death of a very close family member or a loved one. And we know this feeling of loss will be with us for quite a while, until one day, we notice the absence of this feeling.

Then there is the feeling that comes when we know that we are lost, be it because we do not know where we can find the information that will help us to finding something or someone that will help us with our feeling of being lost, such as the experience of being lost in an addiction, when we don't know how to find the answer, how to get rid of the addiction, so that we can return to a better way of living.

And then there is that feeling when we know that we are lost in an unfamiliar place such as in the woods, in a foreign country, or in a big city.

Now that you have an idea of the feeling that I am trying to tell you about, you will understand the feeling of being lost that I had when I was trapped in alcohol addiction and in drug addiction. I knew I was in a life or death situation because of these addictions and I could feel that I was lost. I knew that I did not have the answer for getting out of this problem and it was only when I made that decision to cry out to ***IT*** for help because I knew that I could not do it alone, that I started to become more aware that ***IT*** did listen to me. Since then, in my experience with ***IT***, I have found that ***IT*** does not talk to me directly, and ***IT*** will not send me written messages as to what I should be doing. What I have noticed is that ***IT*** pushes me in the direction that I should be going, in a very peaceful and gentle way.

I have heard from some other people that their relationship with ***IT*** as if ***IT*** were dancing with them, and with others, ***IT*** plays with

them; but with me, ***IT*** pushes me into certain directions. Let me explain this. I no longer try to do things stemming from a personal desire, that arises from my "wants" department, which I have kept as much as I can under lock and key, so that my mind will not use any of what it wants in terms of personal desires. I am now able to recognize these very easily because they are thoughts that are normally expressed in these words: **I want this, and I want that, or this should be done my way.**

You might understand this better if I tell you that since I recognize that ***IT*** supplies my basic needs from the moment I wake up, without having to go to work for anyone other than ***IT***, I just do what comes as my next particular moment. Each day I start out after I have a few cups of my half and half coffee, I then go over to my computer and read my e-mails, and see who needs my attention. I try to answer as soon as possible, for this is one way that ***IT*** pushes me in a certain direction, and oddly enough this has been a very interesting way that ***IT*** pushes me, first, because I have to continue practicing my un-artistic way of writing (my hieroglyphics) to answer these e-mails, and second, because I also have to continue writing because of having to find the words to finish this book, as I send each piece to the ghost writer, and to answer email from people who submit business proposals that will generate monetary energy, for I know that ***IT*** is giving me the opportunity to raise the monetary energy that I will need to pay for the ghost writer, print this book, and enough to promote it through advertising so that it could reach you. I am grateful for the knowledge I have received through writing this book about ***IT***, where I encountered information that I did not have before. I would like to clarify that I really do not need more money because I have the necessary amount to eat for a whole month or more, for my diet has a fixed daily calorie intake. Adhering to this diet allows me to feast at Sizzler's once a week, and even twice a week, if the restaurant would change the buffet menu. As for the rest of my necessities, I can see that ***IT*** has sent me enough monetary energy to meet these, and I try to use my accumulated clothing for I have told myself that I should not buy anything other than new Mac Gregor's (sneakers) at K-mart for $20, which last a long time,

perhaps 10,000 miles (just kidding). I also have to purchase new underwear and socks from time to time since my washing machine gets temperamental. For some reason it gets angry at me and stretches my underwear so badly that I am unable to wear it lest the force of gravity cause them to drop down to my ankles! Sometimes instead of my underwear, my socks put on a disappearing act. But I have solved that problem by buying various pairs of all white socks and now, when there are socks missing I can continue to make matches with those that remain. When I see that I no longer have underwear for 2 weeks, I know it's time to buy some more. So the only new shoes that purchase on a regular basis are my soft $20 Mac Gregor's sneakers, which I buy a half size larger, so that I can insert a cushion insole for extra comfort. One of the advantages of these sneakers is that they are washable. Luckily the washing machine has not found a way to steal or damage them, and after I have gotten my full mileage from them, I then pass them on to my dogs as toys for them to play, which can last them from weeks to months.

As for my other needs I do not worry about them, because the only thing I do have to be concerned about is taking care of my physical body, and making sure that I do not physically damage my brain. I exercise my mind by continuing to absorb more information about ***IT***. This is more important to me than what I used to do before, namely, giving more attention to what so-called "people of importance" were doing. I used to get involved with their actions, judging what they were doing as good or bad. Now I just watch how ***IT*** uses ***ITS*** weight, in terms of how ***IT*** exists out there throughout this planet, and I keep my eyes and ears open to what little we receive as information on what ***ITS*** weight is doing inside ***ITS*** nothingness, as what is happening inside of this Universe as ***ITSELF.***

I also pay attention to the things that I am supposed to do for myself since I am now living alone. I know that I am not supposed to seek some particular person as a mate, for I know that if I go on my mind's desire to find someone, I could end up sick or die. I am at peace knowing that when ***IT*** wants me to have another female

mate, ***IT*** will send it, and we will both know ***IT***, and this process will have to occur within this same living existing moment . If it does not happen, then I only have to continue enjoying what ***IT*** has given me, for I am not interested in a mate for a short term relationship.

So, getting back to my daily schedule, after I answer my e-mail, this sometimes affects some of the things that I would do during this period that I am awake. I say period for I do not live subject to the calendar, in terms of Monday thru Sunday. I just simply exist as one moment of ***ITS*** existence, for I get up when I have rested, and I go to sleep when I see I can no longer keep my eyes open.

Since I really do not have to do anything, I start out my early hours by preparing for my meals, by selecting what flavor beans will I eat, and I prepare my yogurt for dessert, for later on, when I get hungry and I am more tired, I will not have to start to prepare my meal, for I know that it is just 4 minutes away, because that is just how long it will take to warm up my food in the microwave. After that, on certain days do my exercises.

Now, during the day I am aware that I have to keep my eyes and ears open to how ***IT*** will communicate with me, through the people that knock on my door and through the people that call me on the telephone, or through the e-mails that I receive, for I know that these are some of the ways that ***IT*** communicates with me as ***ITSELF.*** Because the truth is that I really do not know what ***IT*** will be doing next as ***IT*** reshapes as you and I and everything else that ***IT*** exists as in terms of ***ITS*** weight. And if there is nothing for me to do in terms of work, I do nothing, as a gift from ***IT*** to me.

I do have certain days that I go to the city to check my P.O. Box, for ***IT,*** as my present moment of existence has me living in a country environment, where I do not get mail service, and while I am down in the city I do some grocery shopping, visit my city friends, and I also use this trip to the city to visit a Sizzler restaurant.

Anyway, I can only see some of the possibilities that ***IT*** might allow me to participate in, as what ***IT*** might reshape ***ITS*** weight into, as events that might happen from what ***IT*** has already sent me and what things might happen from my e-mails and as the effects of what may happen as a result of this book.

I use the words peaceful and gentle to describe the way that ***IT*** uses to make me go in the direction that ***IT*** wants me to go, because I can see that ***IT*** can force us to do things that we have to do in an unpleasant rather than peaceful way. This happens when we refuse to do what must be done. In times such as these, people usually ask the rhetorical question: O God why are you doing this to me?

So now I no longer have the feeling of being lost, for I know that all things are ***IT*** as ***ITS*** heated weight, for I know that I exist within ***IT***, and the feeling that I do get from being close to ***IT***, as ***IT***, has been very durable or long lasting, for I know that this feeling is not based on time, but rather as just one continuous moment of ***ITS*** existence.

And I am no longer alone in our society, for I know that ***IT*** is everything that exists, both as society and everything in it, and I know that as long as I want to be with ***IT*** I will never be alone, for ***IT*** is everything, including the people, places, and things that make me feel secure.

I am no longer searching for anything for I have not lost anything, for everything is ***IT***, and all I have to do is keep on focusing on ***IT***, and the things that ***IT*** wants me to do in a peaceful way.

I no longer wonder what it is that I should be doing in life, for all I have to do is be aware of what ***IT*** wants me to do with ***ITS*** weight, be it in terms of what we call "property" or as the monetary energy. I just have to thank ***IT*** for giving me everything necessary for me to feel well, sleep well, and eat well, and I also thank ***IT*** for allowing me to participate in ***ITS*** reshaping in terms of the processes and actions that I could take part in as a very peaceful

moment of ***ITS*** existence.

So like I have said it is not easy to portray true feelings on paper, For I know that ***IT*** will bring me what I am suppose to have, both in terms of things and people, through my environment, so again thank you (***IT***) and to you, the reader. However, please do not lock on to me. Lock on to ***IT***, for ***IT*** is the best of the best in this place called the Universe.

And as always, if any reader would like more information, please e-mail me at **omnipresentrr@hotmail.com** and I will try and help you. Thank you. Ric-Ricardo

These few pages are dedicated to the friends that I have lost when either I or they moved away. If any reader knows these friends, please ask them if they know me. I miss them and I'd love to reestablish contact!

1 -Bill Gunn, the director, last seen in Nyack, New York
2- Dr. Lee More, Puerto Rico
3 -Dr. Jeremy Villano, Brooklyn
4- Gus and Rita Soto, Syracuse, New York
5- Jules in Puerto Rico
6 -Rachael from the Lower East Side-last seen at E2nd St., Manhattan, NY
7-Emmerly Fox at 14th St., Manhattan NY
8 -Willy Colón, Old San Juan, Puerto Rico
9- Raffy Landrón Old San Juan, Puerto Rico
10-Peto Colón-- the professor in philosophy, Old San Juan, Puerto Rico
11-Jo Ellen and Anita Matos- Old San Juan, Puerto Rico
11-Raymond Ruiz from Stone Ave. - in Jersey City, NJ
12 Edwin Medina, Stone Ave, Brooklyn NY
13-Ramona Hernández, sister to Mina, Liana, and Edwin Reyes - Brooklyn NY

Published by R&R publisher
P.O.Box 9944
Plaza Carolina Station,
Carolina, Puerto Rico 00988-9944

Here are some of the subjects that are covered in volume #2:
Bring index from volume #1 here then remove this line

Este libro se terminó de imprimir
en abril de 2007
en los talleres gráficos de
First Book Publishing of P.R.
Tel. (787) 757-4020